INFORMATION & ON-LINE DATA IN ASTRONOMY

ASTROPHYSICS AND SPACE SCIENCE LIBRARY

VOLUME 203

INFORMATION & ON-LINE DATA IN ASTRONOMY

Edited by

DANIEL EGRET

Observatoire Astronomique de Strasbourg, France

and

MIGUEL A. ALBRECHT

*European Southern Observatory,
Garching bei München, Germany*

SPRINGER SCIENCE+BUSINESS MEDIA, B.V.

Library of Congress Cataloging-in-Publication Data

Information & on-line data in astronomy / edited by Daniel Egret &
    Miguel A. Albrecht.
        p.   cm. -- (Astrophysics and space science library ; v. 203)
    Includes index.
    ISBN 978-94-010-4178-2   ISBN 978-94-011-0397-8 (eBook)
    DOI 10.1007/978-94-011-0397-8
    1. Astronomy--Data processing. 2. Astronomy--Information
services. 3. On-line data processing. 4. Information and retrieval
systems--Astronomy.   I. Egret, D. (Daniel) II. Albrecht, Miguel A.
III. Series.
QB51.3.E43I54 1995
025.06'52--dc20   95-34050

ISBN 978-94-010-4178-2

Printed on acid-free paper

Table of Contents

Preface

THE CHALLENGES OF INFORMATION HANDLING

Astronomy has entered into the era of Great Observatories: the increasing efficiency of cameras and photon collecting devices used by astronomers, either from ground-based telescopes, or from space experiments, is generating an unprecedented accumulation of data. The ability of storing, managing, and giving access to this huge quantity of data, and the associated documents, is one of the major challenges of our science (and of natural sciences in general) for the next decade.

In the 80's, dedicated computer archives and databases, accessible remotely (when the networks allowed), were the appropriate answer to the data retrieval problem. Later on, in order to face the diversity and complexity of access to data for the astronomers as part of their research work, integrated information systems started to build up gradually: examples are the ADS (chapter 17) in the U.S. and the European Space Information System (ESIS; chapter 18) in Europe.

A key lesson from these experiences is that the data have to reside and be maintained at the same place where the expertise is located: *i.e.* as close as possible to the data providers, and generally first users, who are able to understand and process the raw data, and provide the routines for producing final data in physical units — a process which sometimes implies several years of iterative improvements.

However, the scientific teams are generally reluctant to devote time, manpower and money in the different aspects of data distribution to a wider community (documentation, homogenization, building of friendly user interfaces, *etc.*).

At the same time, the growing complexity of data systems, implies a change of concepts. The scientist has not only to manipulate data, but information: observational data, but also documentation about the data, knowledge about the instruments, calibrations, bibliographical references of published results, cross-references to other datasets, *etc.* Pointers and links between these different pieces of the same puzzle have to be constructed.

The World-Wide Web

The recent developments of the World-Wide Web have brought interesting answers to these problems, by providing altogether

▷ easy ways to make all kinds of data and information available through the network (at the level of a group, a department, an institute, or a whole agency);

▷ easy ways to bring together data collected at different physical locations, provided they all use the WWW http protocol;

▷ easy ways to create links between data tables, publications, images or datasets;

▷ easy ways, for a user, to gain access to this variety of data, documentation, and multimedia information: namely NCSA Mosaic, Netscape and other WWW browsers.

The astronomical community showed a very early interest to the World Wide Web and several dozen WWW servers were already available in mid-1993, several hundreds in mid-1994, one thousand at the end of 1994. The astronomical community was ready to jump onto the Web, because of its familiarity with the international collaboration through the data networks, and its computer infrastructure.

A very good way to observe this incredible explosion of astronomical WWW servers is to connect to the AstroWeb [→2] (see also page 201), and browse through the different categories: observing resources, data resources, publication-related resources, people-related resources, organizations, software resources, etc. (a list of the AstroWeb categories, valid for February 1995, is given as Figure 0–1).

Using your favorite WWW browser, you can display the corresponding lists of resources, find a title and eventually a brief summary of the resource: click here and you are witnessing the scientific activity of an astronomical department in Finland; click there and you find the observing schedule on Mauna Kea...

The result of such "Web surfing" may be surprising, exciting, or exotic. It may also be disappointing or upsetting when the "work in progress" sign regularly appears on each and every page, and when promising anchors point to empty pages. This is probably the danger of such an explosion: there is the risk to be flooded with documents of low interest. In the published scientific literature there is a way to solve that: namely the refereeing process. It brings the guaranty that at least somebody else has reviewed the document, and had the opportunity to make a critical review, and to suggest improvements. This is not the case with most of the documents on the Web.

Obviously, the astronomical community will have to organize itself in order to answer properly this challenge, and derive the necessary rule, etiquette, and procedures, for ensuring the highest possible quality standards for on-line information.

THIS BOOK

The current book reflects the important changes mentioned above. When we edited the book "Databases & On-line Data in Astronomy" (M. A. Albrecht & D. Egret, Eds, Kluwer Acad. Publ. 1991) we had envisaged the need of editing an updated version a few years later.

However when the time of volume 2 (the present book) arrived, it was obvious that the technological changes imposed changes in the format of the book (it is now accompanied by an on-line hypertext complementary version) as well as in the title: Information handling is now becoming increasingly important, together with data handling, in the scientific process.

The selection of data holdings and information systems presented here, cannot pretend to be exhaustive. It should be noted that we made a review of the of services presented in the first volume, and those of which the scientific contents have not dramatically changed, are not present in this volume. The reader is asked to refer in this case to the first volume. The updated information about access to these services is available in the on-line version.

The on-line complement of the book includes summary descriptions of the data systems presented here (as well as those presented in "Databases & On-line Data in Astronomy"), together with the complete access pointers (these will be kept up-to-date) and, when available, ready-to-use hyperlinks to the original data systems (either WWW, ftp or telnet services).

This on-line service is available at the following address (URL):

http://cdsweb.u-strasbg.fr/data-online.html

In this book, the reader will also find, at the end of most of the chapters, a section called "Access Pointers" which gives WWW, ftp or telnet addresses for the decribed services.

The book starts with a general discussion of the on-going information (r)evolution. The next chapters present data archives from a number of current space or ground-based experiments (from Compton GRO to COBE). The book continues with chapters presenting the status of several databases (NED, LEDA, BDA), data centers, facilities and information systems (HEASARC, CDS, ADS, ESIS), yellow-page systems (Star*s Family), and library information services.

The last chapters deal with important related issues: how to make information available on the network, networking and information retrieval, data storage technology, and a brief description of the convention used by NED and SIMBAD for bibliographic reference coding.

The full list of authors with their affiliation and address, a table of acronyms, and a subject index can be found at the end.

This work includes a series of clear and up-to-date descriptions of available data systems world-wide, together with their scientific context and motivations. It would not have been possible without the very kind contributions of the authors of the individual chapters, mostly key persons from each project: thanks again to all of them for their enthusiasm and efficiency.

This book should help the reader (astronomer, librarian, computer engineer) to trace his/her way among the jungle of existing on-line astronomical services. It is also meant to encourage the creation of new information services, thus broadening the capacity of the scientific community to observe and understand the universe.

Access Pointers

[→1] On-line complement to this book:
   http://cdsweb.u-strasbg.fr/data-online.html
[→2] AstroWeb: http://fits.cv.nrao.edu/www/astronomy.html

Strasbourg and Garching
February 1995

AstroWeb: Astronomy on the Internet

The AstroWeb database currently contains 1272 distinct resource records:

- **Organizations**
  - O Astronomy Departments (223 records)
  - O Astronomical Societies (14 records)
  - O Space Agencies (40 records)
- **Observing resources**
  - O Telescopes (98 records)
  - O Telescope observing schedules (23 records)
  - O Metereological information (13 records)
  - O Astronomical survey projects (16 records)
- **Data Resources**
  - O Data and Archive Centers (80 records)
  - O Astronomy Information Systems (33 records)
- **Publication-related Resources**
  - O Astronomy & astrophysics preprints & abstracts (60 records)
  - O Abstracts of Astronomical Publications (42 records)
  - O Full-texts of Astronomical Publications (43 records)
  - O Astronomical Bibliographical Services (21 records)
  - O Astronomy-related Libraries (16 records)
- **People-related Resources**
  - O Personal Web pages (347 records)
  - O Jobs (5 records)
  - O Conferences and Meetings (25 records)
  - O Newsgroups (at least somewhat astronomy-related) (18 records)
  - O Mailing Lists (at least somewhat astronomy-related) (2 records)
- **Software Resources**
  - O Astronomy software servers (88 records)
- **Research areas of Astronomy**
  - O Radio Astronomy (42 records)
  - O Optical Astronomy (113 records)
  - O Space Astronomy (87 records)
  - O Solar Astronomy (27 records)
  - O Planetary Astronomy (13 records)
- **Miscellaneous Resources**
  - O Lists of astronomy resources (56 records)
  - O "Pretty Pictures" (61 records)
  - O Miscellaneous Resources (37 records)

This page was computed 22 Feb 1995

Figure 0–1: The categories of the AstroWeb database of Astronomy and Astrophysics resources on the Internet (February 1995).

1

Facets and Challenges of the Information Technology Evolution

André Heck

1–1   Introduction

Our ultimate aim as astronomers or space scientists is to contribute to a better understanding of the universe and consequently to a better comprehension of the place and rôle of man in it. To this purpose, together with theoretical studies, we carry out observations to obtain data that will undergo treatments and studies leading to the publication of results[1]. The whole procedure can include several internal iterations or interactions between the various steps as well as with external fields (instrumental technology, *etc.*), scientific disciplines and information handling methodologies. The trend is also clearly towards panchromatic astronomy as opposed to *'photonic provincialism'* (Wells, 1992).

In the following pages, the concept of *'information'* will cover the observational material, the more or less reduced data extracted from it, the scientific results, as well as the accessory material used by the scientists in their work (bibliographical resources, increasingly important yellow-page services, and so on).

It is common to speak nowadays of the *'information technology (IT) revolution'*. We prefer the more realistic concept of *'IT evolution'*, as we do not know whether or when the process will stop. As far as communication is concerned, many consider we are currently living in a period which is as important for mankind as the XVth century that saw Gutenberg's[2] invention of the movable-type printing process. In a recent book, Peter F. Drucker (1993) predicted a power shift from the entities with financial resources towards

---

[1] Publication is to be understood here as *public announcement* (Webster, 1976). No implicit assumption is made as to the medium used.

[2] Johannes Gensfleisch zur Laden, alias Gutenberg (1400?-1468) lived essentially in Mainz, but spent quite a few years in Strasbourg.

D. Egret and M. A. Albrecht (eds.), Information & On-Line Data in Astronomy, 1–14.
© 1995 *Kluwer Academic Publishers.*

persons or organizations who will have, not necessarily the knowledge it-self, but who will know how to access it and to handle it. At a shorter term, financial wizard George Soros declared in his keynote address at a recent Internet Society conference that he considered the current connectivity a crit-ical component for the Open Society which was the basis for political and economical stability as well as organizational success and self-fulfillment in the XXIst century (Rutkowski, 1994, [→1]).

The previous reference itself is a good example of the changes occurring nowadays in communication: it is indeed pointing, not to a classical con-tribution on paper, but to an electronic document available through new tools and, last but not least, shortly on-line after the *Networld+Interop '94* conference in Tokyo where it had been given as a keynote address on July 29, 1994.

1–2   The IT evolution

Compared with the book *Databases & On–line data in Astronomy* (Albrecht & Egret, 1991), the present volume is a striking example of this evolution. Refer also to Heck & Murtagh (1993) for a review on advanced information retrieval tools with special emphasis on astronomy and related space sciences. The ADASS (Worrall *et al.*, 1992; Hanisch *et al.*, 1993; Crabtree *et al.*, 1994) and ALD (Murtagh & Heck, 1988; Heck & Murtagh, 1992) conference series also testify *i.a.* the dramatic progress in computer capabilities, data reduction, communications, desktop and electronic publishing, and, more generally, information handling. The substance of libraries and the rôle of librarians are also significantly changing. The distinction between the formal and the so-called 'grey' literature is everyday more difficult to make, as both of them are taking similar electronic shapes. The formal 'invited papers' are becoming multimedia presentations.

The structure of information itself is becoming different: beyond the classical quasi-linear layout of publications on paper, there are more and more frequently sets of *documents* with hypertext (see *e.g.* Nielsen, 1990 & Landow, 1992) links, the structure of which can be closer adjusted to the own mental structure of the authors. Too much time has often been wasted in the past when authors had to get all their bits of ideas and side notes into a smooth flow for a 'classical paper'. Some of them could not actually manage that entirely and this was specially perceptible in papers full of footnotes or with long series of notes in appendices, some of them referring to each others in hierarchical or open loops. It would have been actually easier to prepare this very paper in hypertext structure even if modern text processors allow to move around pieces of text at will. Writing in a linear way after practicing hypertext has become almost as difficult as hand-writing an article or a book on sheets of paper after having enjoyed the flexibility of computerized text-processing systems.

The information material as a whole is now existing in an increasingly distributed way. Mankind should now definitively be protected from the consequences of a disaster such as the burnout of Alexandria's libraries in the third century AD. If data centres have seen their rôle evolving and if they tend now to act more as a hub towards distributed specialized repositories of different types of data and material (rather than, as in the past, holding as much as possible themselves and carrying out the integration work on their very location), this is largely due to the fact that the IT evolution has brought in major modifications as to hardware and connectivity as well as new tools (client/server facilities, resource discovery packages, *etc.*) and concepts (hypertext/hypermedia concepts, virtual libraries, *etc.*).

From its origin as part of a US government project (ARPA research project on inter-networking in the early 1970s), the Internet has become the major network linking nowadays millions of machines and tens of millions of people around the planet. Today, little of the Internet is controlled directly by governmental bodies and there are already impressive listings of commercial companies on the 'Web' [→1]. It is obvious that an estimated market of 20-30 millions people on Internet (Rutkowski, 1994) is particularly appealing!

Developed at CERN (see *e.g.* Berners-Lee *et al.*, 1992 & 1994, White, 1993), the World-Wide Web (WWW) actually allowed for the first time Internet users to reach the various types of services accessible via this network (ftp, gopher, telnet, *etc.*) through a consistent and abstract mechanism. Based on a hypertextual language (HTML – see Berners-Lee & Connolly, 1993), WWW calls on a specific protocol (HTTP – see *e.g.* Berners-Lee, 1994a) for a rapid transfer mechanism based also on Uniform Resource Locators (URLs) providing the electronic addresses of documents accessible on the web (see *e.g.* Berners-Lee, 1994b and [→6]). These URLs are already heavily used as references in this chapter. More flexibility (figures, tables, forms, footnotes, margins, additional special characters, *etc.*) will result from the use of future features of HTML (see *e.g.* Raggett, 1994 and [→2]).

The development of tools such as Mosaic (see *e.g.* Hardin, 1993 and [→3]) at NCSA made it much easier to move ('navigate') happily from one document to another one through virtual libraries on the web, as well as to give access to already existing services (such as the CDS services —see chapter 16, and other chapters in this book). The distributed information is there (see *e.g.* Christian & Murray, 1994, Egret & Heck, 1994, and Fullton, 1994) and it is up to us to make the best use of it at the maximum of its current and future possibilities. More generally speaking, each document or 'page' connecting itself on the web contributes to the enlargement of the virtual libraries that become everyday closer to the original Ted Nelson (1981)'s vision of a virtual encyclopaedic library (Xanadu project).

The publications of the *Association for Computational Machinery (ACM)* and the proceedings of the meetings organized by its various Special Interest Groups (SIGs) are precious sources of information on the IT evolution and the various related issues. The meetings sponsored by the ACM's *Special Interest Group on Hypertext and Hypermedia (SIGLINK)* and *Special Interest Group on Information Retrieval (SIGIR)* are particularly recommended.

1–3   Think about it twice

This section will be devoted to restate a few commonly misunderstood points, beginning with electronic publishing. First, there is still a confusion often made between desktop and electronic publishing. The former one can be understood as a way of producing locally through relatively sophisticated software packages and laser printers high-quality printed material ready for reproduction by a publisher with the traditional camera-ready technique. The latter term would rather concern the electronic submission of material straight to the publisher who will work directly on the electronic files and get the paper, journal, or book ready for being printed through a succession of computer-assisted steps.

This is what was considered as the state of the art at the first international astronomical colloquium on electronic publishing (Heck, 1992). However the popularization of hypertextual structures has added new degrees of freedom. Unfortunately many people nowadays have still too classical an approach, seeing nothing more in electronic documents than an electronic version of a traditional publication on paper. This leads to semantically conflictual expressions such as 'electronic preprints'! It is also interesting to note that a journal on paper has been launched a few years ago by Wiley with the title *Electronic Publishing* while other commercial publishers are already heading at full steam to take advantage of the new hypertext techniques (Dougherty, 1993).

To the frequent question *'Do we still need the classical publishers?'*, our answer is *'Yes'*, simply because we still see the new IT media as complementary of the traditional ones. Of course, there are quite a few valuable arguments raised against it (see *e.g.* Berners-Lee, 1992 & Heck, 1992), especially at a time when library costs are spiraling upwards as opposed to the electronic material where the costs are widely distributed and where there is apparently a reduction of the overall invested energy and manpower. There are also quite a few arguments in favor as the expertise (and the means) by commercial publishers to protect copyrighted works. But how often actually had this procedure to be applied in the past and do we really need it in the future?

It is a fact that creators, authors, contributors, and so on, are worried by the fragility of their work under electronic format (making illicit copying easier, etc. – in this respect, see *e.g.* Samuelson, 1994), but this fear is basically linked to the aspect of recognition. This, and more generally the ethical behavior, will have to be adapted to the new communication practices (see below). It is probably unfair to consider as competitive the classical journals (thus the traditional libraries) with electronic information (thus with the virtual libraries), since the hypertextual structure of the latter, its distributed nature as well as its various digitized media make it primarily complementary of the former ones. One must be careful before burying too quickly the 'classical' publishers, the traditional journals and our beloved librarians (although the latter ones will actually become more and more electronic-information specialists).

The CD-ROM is another case of a sometimes misunderstood technology. Its biggest advantage is its compactness and the following example is well known: the specification of the the F-18 aircraft required some 300,000 pages of documentation which is equivalent to a storage volume of about 68 cubic feet of paper. This could be put on a 0.04 cubic feet CD-ROM. But this medium is a frozen repository of information compare to permanently updated databases available on-line. And it requires quite an exercise to be consulted: to switch on the computer, to get the CD drive running with the *ad hoc* software, and so on, not to speak of standards. Who has already used a CD-ROM on a laptop or notebook on a plane? Again here, it is important to stress that the CD-ROM is a complementary technique, and not a panacea aimed at replacing indiscriminately the classical documentation.

It is actually surprising that new journals are being issued on this medium when connecting facilities are multiplying everywhere, including the wireless ones. For classical journals structured along the traditional quasi-linear layout, the eyes and the hands remain particularly efficient tools for scanning them. This is why, for limited queries, people still often prefer searching through journals or books, rather than switching on computers, sliding in CD-ROMs or floppies, and keying codes and identifiers. This is not necessarily guided only by habits and inertia, but more often motivated by better efficiency and handiness. See for instance the experiment made in France with the on-line telephone book[3].

---

[3]When the French telephone company introduced the Teletel/Minitel service, a cutting-edge technology at that time providing each telephone set with a video terminal, it was thought that the yearly telephone books on paper would disappear since the first service offered free of charge through Teletel was the search for telephone numbers in a permanently updated on-line directory through a relatively friendly system. Nowadays Teletel technique has become an enormous success and has been largely exported, offering access to thousands of services ranking from databases, games, theater bookings, bank account management to mail ordering and so-called 'blue' ('pink' in French) mailers, but the telephone books are

In a very interesting essay, Moore (1991) investigated the adoption life cycle of new technologies, pictured them as bell-shaped curves divided in zones corresponding to various behaviors (innovators, early adopters, early majority, late majority, and laggards), and pointed out cracks between these zones. The largest one, the 'chasm', appears to separate the early adopters from the early majority. This is where hypertextual techniques[4] are currently emerging from and where other techniques such as pen computing (see *e.g.* Meyrowitz, 1992), virtual reality (see *e.g.* Rheingold, 1991) and others are still waiting for better times.

Thus new technologies are not always immediately adopted and this adoption does not necessarily undergo a smooth process. It is also interesting to see how some tools or products laboriously and expensively developed for solving urgent needs at a specific time are then used much more successfully by a subsequent technology. In a keynote address at *ECHT '92* , Ritchie (1992) brilliantly explained the remarkable technological achievements made at the canal age to bring water and barges over bridges and through tunnels because of the then pressing needs for transporting ore and other goods. All this technology was subsequently much more efficiently exploited by the railway.

1–4   Yellow-page services

The concept of *yellow-page services* was probably publicly introduced for the first time in our community at the *ALD-II Conference* in Haguenau (Heck & Murtagh, 1992). In the volume by Albrecht & Egret (1991), the chapter entitled *Astronomical Directories* (Heck, 1991) was essentially dealing with classical publications on paper: directories of organizations, lists of electronic addresses, and so on. Refer to that paper for a historical review on earlier directories. Some files started to be retrievable electronically (*e.g.* Benn & Martin's, 1990, lists of e-mail addresses). In the section (*Future Trends*) of that chapter, we were looking forward to using on-line and networking facilities that have become a daily reality since.

Meanwhile indeed the existing references products have tuned them-selves to the new capabilities brought in by the IT evolution. New reference products naturally derived from the new media available. In a separate chapter (Heck, 1995, page 195) and in the references quoted therein, we describe a comprehensive set of yellow-page services: the Star*s Family.

---

still there and one can even find now directories on paper of . . . Teletel/Minitel codes!

[4]Hypertext – a term coined by Nelson (1967) – has been around for quite some time already. See *e.g.* Smith & Weiss (1988) and the subsequent papers of that special issue of the *ACM Communications* on hypertext they edited.

To help users to find their way *'in the jungle of astronomical on-line data services'* (Egret, 1994), several large compilations of resources have already been made available: refer *e.g.* to Feigelson & Murtagh (1993) for public software, to Davenhall (1993) for on-line bibliographic and information services, as well as to Andernach *et al.* (1994) for exhaustive lists of resources of all kinds in astronomy and related space sciences. It is important to keep such databases of resources on-line and to update them as frequently as possible, especially because of the pace of evolution is difficult to follow, and even more for the non-specialist.

AstroWeb (Jackson *et al.*, 1994) is a new WWW resource providing URLs to astronomy resources available on Internet (see, in this book, page 201).

We can only encourage organizations (institutions, associations, funding agencies, and so on), as well as individuals, to set their own resources reachable on WWW, bearing in mind that they become 'public' *ipso facto* and that all documents are then 'published'. The question of validation (for instance via a refereeing system) is a distinct one to be addressed independently, possibly also through electronic means (Heck, 1992).

1–5   Maintenance and quality

The maintenance process of information resources must be continuously improved from lessons learned with time and by using the most adequate tools. Refer to the chapter on the *Star**s Family* in this volume (page 195) for a few considerations specific to this case. Generally speaking, information has to be collected, verified, de-biased, homogenized, and made available not only in an efficient way, but also through operationally reliable means (it becomes useless if plugged into a confidential network or reachable through deficient routers). Redundancies have to be avoided; precision is and details can be extremely important.   Last but not least, the continuous political evolution of the world has also to be taken into account and one must be permanently in alert on practical aspects such as restructurings of postal codes, telephone area codes, and so on.

Professional file construction techniques are nowadays mandatory. These however cannot save the extensive background, unrewarding and very careful work which is indispensable for the compilation of a valuable resource. One could never stress enough the importance of this obscure daily work consisting of patiently collecting data, checking information and updating the master files. If scientists have a natural tendency to design projects and software packages involving the most advanced techniques and tools, there is in general less enthusiasm for the painstaking and meticulous long-term maintenance which builds up however the real substance of the resources.

This has also to be carried out by knowledgeable scientists or documentalists and cannot be delegated to unexperienced clerks.

Efficient working procedures have to be worked out and may vary from one person to another one depending of working abilities, intellectual structure and education, and so on. Once again, maintaining a database is a very demanding tedious task, a time-consuming and endless everyday occupation requiring memory, aesthetical feeling, as well as pragmatism and acumen towards current and future needs of users.

Information retrieval *per se* is raising a number of evaluation issues[5]. The fashion is now shifting towards designing and experimenting quality control processes. This might be a very serious matter or a big joke. Until further evidence is brought up, we believe that the best quality assurance (accuracy, homogeneity, consistency exhaustivity, *etc.*) has to be achieved when collecting and entering the data themselves. None of the algorithms currently available has really convinced us of their absolute necessity and satisfactory efficiency. Again here, developing such processes is an appealing challenge for scientists, but most of the algorithms designed work statistically. For a database user, it does not matter much whether the material queried is accurate up to 95% or 98%. The user wants to find the piece of information he/she is looking for, and, if found, this has to be accurate. If what he/she is looking for is not available or is wrong, this is what will be remembered. All these considerations are obvious if a telephone book or database is taken as a model for yellow-page services.

1–6   Plenty of work on hand

This explosion of documents on the web is not a bed of roses. New facilities and new possibilities bring in naturally new questions and new problems. Some of the Mosaic servers have already reached a quite fair degree of maturity. Others are still a bit in a wild stage by lack of structure and homogeneity or simply because they offer, let us say it frankly, a significant amount of rubbish of little interest. Although quite a few features have been adopted *de facto* by the developers of documents on the web, there will be, sorry, there is, a definite need for a *WWW ethical charter*. It could concern quite a number of features from the substances of the documents themselves to their aesthetical presentation and a number of recommended functionalities.

Each document should be 'signed' in some way and contacts for comments should be provided, if possible involving dynamic e-mail facilities (but please do not forget also to indicate the full name of the organization with its address, phone and fax numbers, and so on; an access map is also

---

[5]See *e.g.* Harman (1992) and the subsequent papers of the corresponding special issue.

quite welcome). The multiplication of large pictures and icons should also be discouraged and limited to what is really essential for the sake of information retrieval and network efficiency. The fashion of putting unrelated material, especially pictures of wives, girlfriends, actors, actresses or top models is an insult to the person querying the pages in hope of some interesting information (or, at the very least, it is a ridiculous waste of time for the user). People should also remember that humour is not always exportable.

Hypertext itself is too often badly used to structure the documents. Pointing to external resources should also be preferred to cutting and pasting in. In any case, proper credit to the material used should always be clearly indicated. Only well-tested documents should be put on line. It is easy to create working directories on which Mosaic (for instance) can be run locally. URLs should not be changed unless absolutely necessary and, in such cases, links should be provided from the obsolete ones to the new ones.

Since tools such as Mosaic make it so easy to download the original files, crediting the sources appropriately becomes critically important. It is actually smarter and more elegant to insert a hyperlink to the original document since it will point then always to the freshest version of the file. This brings us to *security* issues, involving monitoring, restricted access, confidentiality and so on. Away from governmental policies (such as the Clipper chip project in the US that is raising substantial controversy), there is no golden rule on security issues: it is up to each local 'webmaster' to put the appropriate securities according to the material concerned. Some resources require appropriate clearance (password, account number and so on); others will be only partially retrievable in a specific query (such as large copyrighted databases); finally, other documents are freely accessible and usable, conditioned to a minimum of ethical behavior (see above).

Legal aspects (copyright, electronic signatures, *etc.*) are also extremely important and jurists are busy setting up references for the computerized material. Particularly in this case, there might be variations from country to country when the law already exists. However, with the world globalization of electronic communications, one can expect —and hope for— a quick harmonization of the various references and procedures. On such matters, refer to Samuelson (1993) and to her very interesting regular column *'Legally Speaking'* in the *ACM Communications*.

Last, but not least, there are non-negligible *educational aspects* to be taken into account as to the introduction and training of young and not-so-young people to the new technologies within the various communities. This is true not only for scientists, but also for librarians and documentalists who will see their rôle significantly changing within their institution and who will be increasingly dealing with a virtual material.

1–7   A few last comments

It would be dangerous —and pretentious— to play here the game of guessing the long-term impact of the IT evolution. What is sure is that technological progress will play a key rôle in future orientations. But these are still too fuzzy and all predictions are risky. Two years ago, Mosaic was unknown while today it allows a daily cyberspace[6] navigation on a planetary scale. Who would still dare to plan computer technology and information handling a lustrum ahead? It is more than ever time for managers to rely on members of their staff gifted with intuition.

We might have nevertheless to reassess thoroughly the process of *evaluation* applied for financing research and that conditions the need for *recognition* (which is currently based essentially on traditionally validated publications on paper) for getting positions (grants and salaries), acceptance of proposals (leading to data collection), and funding of projects (allowing materialization of ideas).

The whole IT evolution emphasized the need for multidisciplinary approaches and that is where meetings organized by ACM, CODATA[7], SIAM[8] , and others, are valuable forums.

It is also clear that the advent of the web makes interdisciplinary communications much easier, more emulating and more inspirational. One must however be cautious with encyclopaedic tendencies resulting from the IT evolution and, even within a specific scientific discipline, one must refrain from engulfing enormous amounts of energy and manpower in oversized endeavours with questionable return. The history of European astronomy has this pitfall illustrated by the ambitious project of the *Carte du Ciel* that mobilized during critical decades the resources of various countries and delayed the flourishing of astrophysics in Europe while it was happily taking off on the other side of the Atlantic. If the technology has evolved, similar wrong steps might be in the making. *Trahit sua quemque voluptas.*

Acknowledgements

It is a pleasure to acknowledge here the impact of numerous readings and conversations we do not dare detailing in order to avoid forgetting a couple

---

[6]For the literary navigation style, the *'cyberpunk'* school, refer *e.g.* to Gibson (1986 & 1993).

[7]CODATA is a committee of the International Council of Scientific Unions (ICSU) dealing with data of importance to science and technology, their compilation, critical evaluation, retrieval and management.

[8]Society for Industrial and Applied Mathematics.

of them whose contribution, if sometimes not always important in substance, has been nevertheless enlightening.

References

[1] Albrecht, M.A. & Egret, D. 1991, *Databases & on-Line data in astronomy*, Kluwer Acad. Publ., Dordrecht, xiv + 274 pp. (ISBN 0-7923-1247-3)

[2] Andernach, H., Hanisch R.J. & Murtagh F. 1994, Network Resources for Astronomers, *Publ. Astron. Soc. Pacific*, **106**, 1190 (see also: [→11])

[3] Anklesaria, F. & McCahill, M. 1993, *The Internet gopher*, in *Intelligent Information Retrieval: The Case of Astronomy and Related Space Sciences*, eds. A. Heck & F. Murtagh, Kluwer Acad. Publ., Dordrecht, 119-125

[4] Benn, Ch. & Martin, R. 1990, Electronic Mail Guide and Directory, Roy. Greenwich Obs., Herstmonceux (electronic files)

[5] Berners-Lee, T.J. 1992, Electronic publishing and visions of hypertext, *Physics World* **June**, 14-16

[6] Berners-Lee, T.J. 1994a, Hypertext transfer protocol (see also the URL: [→5])

[7] Berners-Lee, T.J. 1994b, WWW names and addresses, URIs, URLs, and URNs (see the URL: [→6]

[8] Berners-Lee, T.J. & Connolly, D. 1993, Hypertext markup language: A representation of textual information and metainformation for retrieval and interchange, *CERN Internal Draft* (see also the URLs: [→4])

[9] Berners-Lee, T.J., Cailliau, R., Groff, J.F. & Pollerman, B. 1992, World-Wide Web: The information universe, in *Electronic Networking: Research, Applications and Policy*, 52-58 (see also [→3])

[10] Berners-Lee, T.J., Cailliau, R., Luotonen, A., Nielsen, H.F. & Secret A. 1994, The World-Wide Web, *Comm. ACM* **37-8**, 76-82

[11] Christian, C.A. & Murray, S.S. 1994, The Earth Data System and the National Information Infrastructure Tested, in *Astronomical Data Analysis Software and Systems III*, eds. Crabtree, D.R., Hanisch, R.J. & Barnes, J., *Astron. Soc. Pacific Conf. Series* **61**, 45-48 (see also: [→9])

[12] Crabtree, D.R., Hanisch, R.J. & Barnes, J. 1994, Astronomical data analysis software and systems III, *Astron. Soc. Pacific Conf. Series* **61**, xxvi + 542 pp. (ISBN 0-937707-80-5)

[13] Davenhall, A.C. 1993, An astronomer's guide to on-line bibliographic databases and information services, Starlink User Note **174.1**

[14] Dougherty, D. 1993. Forging the business of hypertext publishing, Hypertext '93 Tutorial **12**, Seattle, 42 pp.

[15] Drucker, P.F. 1993, Post-capitalist society, Harper Business, New York, 232 pp. (ISBN 0-88730-620-9)

[16] Egret, D. 1994, In the jungle of astronomical on-line data services, in *Astronomical Data Analysis Software and Systems III*, eds. Crabtree, D.R., Hanisch, R.J. & Barnes, J., *Astron. Soc. Pacific Conf. Series* **61**, 13-17

[17] Egret, D. *et al.* 1995, The Strasbourg astronomical Data Center, CDS (this book, page 163)

[18] Egret, D. & Heck, A. 1994, WWW in astronomy and related space sciences, in *Second Internat. WWW Conf.*, in press (see also: [→20])

[19] Feigelson, E.D. & Murtagh, F. 1993, Public Software for the Astronomer: An Overview, *Publ. Astron. Soc. Pacific* **104**, 574-581

[20] Fullton, J. 1993, WAIS, in *Intelligent Information Retrieval: The Case of Astronomy and Related Space Sciences*, eds. A. Heck & F. Murtagh, Kluwer Acad. Publ., Dordrecht, 113-117

[21] Fullton, J. 1994, Distributed astronomical data archives, in *Astronomical Data Analysis Software and Systems III*, eds. Crabtree, D.R., Hanisch, R.J. & Barnes, J., *Astron. Soc. Pacific Conf. Series* **61**, 3-9

[22] Gibson, W. 1986, Neuromancer, Grafton, London, 318 pp. (ISBN 0-586-06645-4)

[23] Gibson, W. 1993, Virtual light, Viking, London, 296 pp. (ISBN 0-670-84890-5)

[24] Hanisch, R.J., Brissenden, R.J.V. & Barnes, J. 1993, Astronomical data analysis software and systems II, *Astron. Soc. Pacific Conf. Series* **52**, xxx + 584 pp. (ISBN 0-937707-71-6)

[25] Hardin, J. 1993, Human collaborations technologies for the Internet - NCSA Mosaic and NCSA Collage, communication to *Astronomical Data Analysis Software and Systems III*, Victoria, BC, 13-15 Oct. 1993

[26] Harman, D. 1992, Evaluation issues in information retrieval, *Information Processing & Management* **28**, 439-440

[27] Heck, A. 1991, Astronomical directories, in *Databases & On-Line Data in Astronomy*, eds. M.A. Albrecht & D. Egret, Kluwer Acad. Publ., Dordrecht, 211-224

[28] Heck, A. 1992, Desktop publishing in astronomy and space sciences, World Scientific, Singapore, xii + 240 pp. (ISBN 981-02-0915-0)

[29] Heck, A. 1995, The Star*s Family: An example of comprehensive yellow-page services (this book, page 195)

[30] Heck, A. & Murtagh, F. 1992, Astronomy from large databases II. Haguenau, 14-16 September 1992, *ESO Conf. & Worshop Proc.* **43**, x + 534 pp. (ISBN 3-923524-47-1)

[31] Heck, A. & Murtagh, F. 1993, Intelligent information retrieval: the case of astronomy and related space sciences, Kluwer Acad. Publ., Dordrecht, iv + 214 pp. (ISBN 0-7923-2295-9)

[32] Jackson, R., Wells, D., Adorf, H.M., Egret, D., Heck, A., Koekemoer, A. & Murtagh, F. 1994, AstroWeb – A database of links to astronomy resources (announcement of a database), *Astron. Astrophys. Suppl.*, **108**, 235 (see also: [→12])

[33] Landow, G.P. 1992, Hypertext: The convergence of contemporary critical theory and technology, Johns Hopkins Univ. Press, Baltimore, xii + 242 pp. (ISBN 0-8018-4280-8)

[34] Meyrowitz, N. 1992, Pen computing: The new mobility of interactive applications, *ECHT '92 Tutorial* **T7**, Milano, 36 pp.

[35] Moore, G.A. 1991, Crossing the chasm, Harper Business, New York, xviii + 224 pp. (ISBN 0-88730-519-9)

[36] Murtagh, F. & Heck, A. 1988, Astronomy from larges databases. Scientific objectives and methodological approaches, *ESO Conf. & Workshop Proc.* **28**, xiv + 512 pp. (ISBN 3-923524-28-5)

[37] Nelson, T.H. 1967, Getting it out of our system, in *Information Retrieval: A Critical Review*, ed. G. Schechter, Thompson Books, Washington, 191-210

[38] Nelson, T.H. 1981, Literary Machines, Mindfull Press

[39] Nielsen, J. 1990, Hypertext and hypermedia, Academic Press, San Diego, xii + 264 pp. (ISBN 0-12-518410-7)

[40] Raggett D. 1994, A review of HTML+ document format, in *First Internat. Conf. on the World-Wide Web*, in press (see also the URL: http://www1.cern.ch/PapersWWW94/dsr.ps)

[41] Rheingold, H. 1991, Virtual reality, Simon & Schuster, New York, 416 pp. (ISBN 0-671-69363-8)

[42] Ritchie, I. 1992, The future of electronic literacy: Will hypertext ever find acceptance?, *ECHT '92 Keynote Address*, Milano & private communication

[43] Rutkowski, A.M. 1994, The present and future of the Internet: Five faces, in *Networld + Interop '94 Conf.*, in press (see also [→1])

[44] Samuelson, P. 1993, Intellectual property protection *Hypertext'93 Tutorial* **21**, Seattle, 66 pp.

[45] Samuelson, P. 1994, Copyright's fair use doctrine and digital data, *ACM Comm.* **37-1**, 21-27

[46] Smith, J.B. & Weiss, S.F. 1988, Hypertext, *ACM Comm.* **31**, 816-819

[47] Webster Third New International Dictionary 1976, Merriam Co., Chicago, lxxxvi + 3136 pp. (ISBN 0-87779-106-6) (three volumes)

[48] Wells, D.C. 1992, Open challenges, in *Astronomy from Large Databases II*, eds. A. Heck & F. Murtagh, *ESO Conf. & Workshop Proc.* **43**, 143-148

[49] White, B. 1993, WorldWideWeb (WWW), in *Intelligent Information Retrieval: The Case of Astronomy and Related Space Sciences*, eds. A. Heck & F. Murtagh, Kluwer Acad. Publ., Dordrecht, 127-133

[50] Worrall, D.M., Biemesdorfer, C. & Barnes, J. 1992, Astronomical data analysis software and systems I, *Astron. Soc. Pacific Conf. Series* **25**, 551 pp. (ISBN 0-937707-77-9)

Access Pointers

[→1] Rutkowski, 1994: http://info.isoc.org/interop-tokyo.html

[→2] Commercial Services Index: http://www.directory.net/

[→3] Berners-Lee *et al.* 1992:
http://info.cern.ch/hypertext/WWW/TheProject.html

[→4] HTML: http://info.cern.ch/hypertext/WWW/MarkUp/MarkUp.html,
ftp://info.cern.ch/pub/www/doc/html-specs.ps,
ftp://info.cern.ch/pub/www/doc/html-specs.txt

[→5] HTTP:
http://info.cern.ch/hypertext/WWW/Protocols/HTTP/HTTP2.html

[→6] URLs:
    http://info.cern.ch/hypertext/WWW/Addressing/Addressing.html
[→7] HTML+:
    http://info.cern.ch/hypertext/WWW/MarkUp/HTMLPlus/htmlplus_1.html
[→8] NCSA Mosaic: http://www.ncsa.uiuc.edu/Pubs/Mosaic/mosaic.html
[→9] Christian and Murray 1994:
    http://niit1.harvard.edu/~carolc/utils/cmsi_adass/section3.1.html,
    http://niit1.harvard.edu/~carolc/utils/cmsi_adass/cmsi_adass.html
[→10] Egret and Heck 1994:
    http://zaphod.ncsa.uiuc.edu/Astronomy/egret/egret.html
[→11] Andernach et al. 1994:
    http://www.hq.eso.org/online-resources-paper/rrn.html
[→12] AstroWeb: http://cdsweb.u-strasbg.fr/astroweb.html,
    http://meteor.anu.edu.au/anton/astronomy.html,
    http://fits.cv.nrao.edu/www/astronomy.html,
    http://ecf.hq.eso.org/astro-resources.html,
    http://stsci.edu/net-resources.html

2

The Compton Gamma-Ray Observatory Archive

Paul Barrett

2–1 Introduction

The *Compton Gamma-Ray Observatory* (henceforth *Compton* Observatory) was initially conceived, designed and developed as a Principal Investigator (PI) class satellite due to the complex nature of its instrumentation and data. The decision to extend the mission's nominal lifetime from three to no less than six years prompted NASA to include *Compton* Observatory in its Great Observatory Program and, hence, to develop a Guest Investigator (GI) Program to maximize its scientific productivity. A key part of the GI Program was the creation of the *Compton* Observatory Science Support Center (COSSC) whose responsibilities include archiving the data from the four instruments, directing the Proposal Peer Review, and assisting GIs in their data analysis with support from the PI teams.

The *Compton* Observatory was designed to study γ-ray emission from the following sources:

▷ pulsars and accreting compact objects such as white dwarfs, neutron stars and black holes,

▷ the decay of radioactive nuclei such as ^{26}Al, which is a by-product of novae, supernovae, and nuclear reactions in the core of massive stars,

▷ the galactic diffuse emission resulting from cosmic-ray nuclei and electrons interacting with matter and photons in the Milky Way galaxy,

▷ radio galaxies and active galactic nuclei such as Seyfert galaxies, BL Lacertae objects, and quasars,

▷ the extragalactic diffuse emission and primordial black holes which are a result of conditions in the early universe,

▷ γ-ray bursts whose origins are unknown,

▷ those sources that have yet to be discovered or identified.

D. Egret and M. A. Albrecht (eds.), Information & On-Line Data in Astronomy, 15–26.
© 1995 *Kluwer Academic Publishers.*

On 16 May 1991, following an $\approx$ 40 day check-out and calibration period, *Compton* Observatory began its Phase 1 observations which lasted about 15 months. During this period, the two wide-field instruments, CompTel and EGRET, performed a nearly uniform all-sky survey. OSSE, with its smaller field-of-view, made $\approx$ 55 high-priority, discrete observations, while BATSE, being an all-sky monitor, performed normal operations. During Phase 2 (beginning 17 November 1992), Guest Investigators were allocated $\approx$ 30% of the observing time: this was increased to $\approx$ 60% for Phase 3 (beginning 15 August 1993). Phase 4 will contain no predetermined amount of observing time for either GIs or PI teams. The first observing period of Phase 4, called Cycle 4, will last $\approx$ 1 yr. Subsequent cycles are expected to last a similar length of time.

2–2   The Observatory

The *Compton* Observatory was launched on April 5, 1991 by the Shuttle *Atlantis* and deployed into a 450 km circular orbit with an inclination of 28.5°. By October 1993 the orbit had decayed to 345 km. A three-stage reboost was successfully conducted during October–December 1993, restoring the observatory to a nearly circular 450 km orbit. There is sufficient on-board propellant for several additional reboosts. The next should not be necessary until about 1999 due to decreasing solar activity. The present low-Earth orbit is intended to minimize the charged-particle background which would increase at a higher altitude, though atmospheric drag would be reduced.

The *Compton* Observatory (Figure 2–1) has a mass of 15,900 kg of which 6000 kg comprises the four distinct, yet complementary, scientific instruments (see Table 2–1). The observatory is three-axis stabilized with an accuracy of 2 arc-minutes and can be slewed to any part of the sky. The spacecraft clock has an absolute accuracy of 0.1 ms. Since the failure of both spacecraft tape recorders in March 1992, data has been telemetered in real-time via NASA's Tracking and Data Relay System Satellites (TDRSS). Initially telemetry was limited to about 65% of the orbit, but with a reconfiguration of the TDRSS system and the establishment of a tracking station in Australia, the telemetry is now greater than 80%. In addition BATSE and OSSE have modified versions of their spacecraft software to enhance data storage and delay telemetry allowing them nearly 100% orbital coverage.

2.1   OSSE

The Oriented Scintillation Spectroscopy Experiment (OSSE) is designed to perform comprehensive observations of astrophysical sources in the 0.1 to 10 MeV range, with some capability above 10 MeV for solar γ-ray and neutron

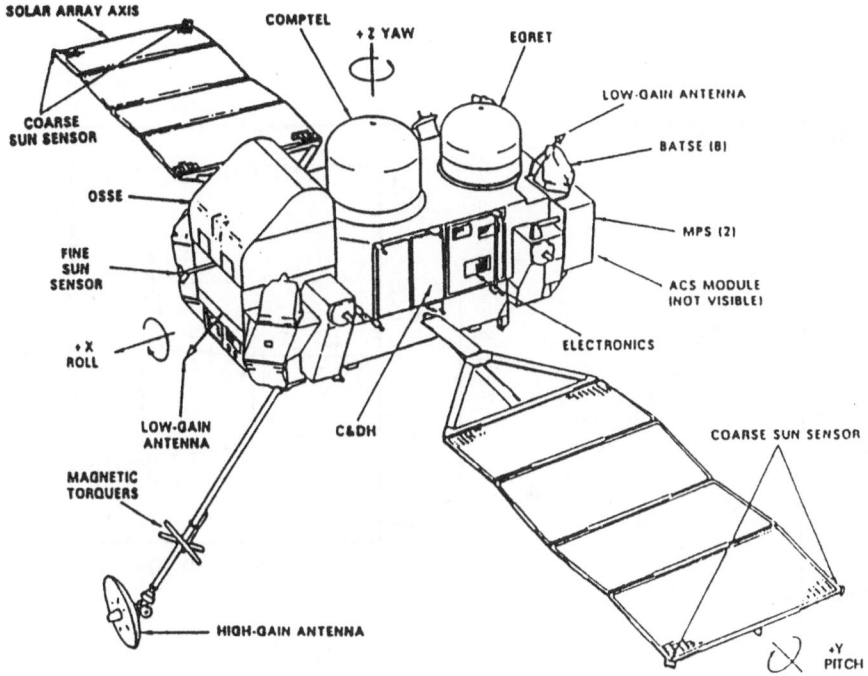

Figure 2–1: The *Compton* Observatory

observations. The detectors are NaI(Tl)-CsI(Tl) phoswiches with a field of view of 3.8° × 11.4°. The energy resolution is approximately 8% at 0.661 MeV. The OSSE detectors are generally operated in co-axial pairs, using an offset pointing capability to switch between on-source and off-source pointings every 2 minutes. While one detector observes the source , the other measures the background. The full detector rotation capability of 192° can be used to observe additional sources (secondary targets) during periods of earth occultation of the primary targets. In addition, OSSE can slew to observe the Sun in response to the detection of a solar flare by BATSE. Normally, OSSE collects individual spectra which are transmitted every 4 s. An alternate data mode provides time resolution of up to 0.125 ms for observations of pulsars. Figure 2–2 shows an OSSE spectrum of the galactic center. The Principal Investigator of OSSE is Dr. J. D. Kurfess of the U.S. Naval Research Laboratory.

2.2 COMPTEL

The Compton Telescope (CompTel) identifies γ-rays in the 1 to 30 MeV range by successive interactions in the telescope: first a Compton collision occurs in one of seven (low-Z) liquid scintillators and then is, idealistically, completely

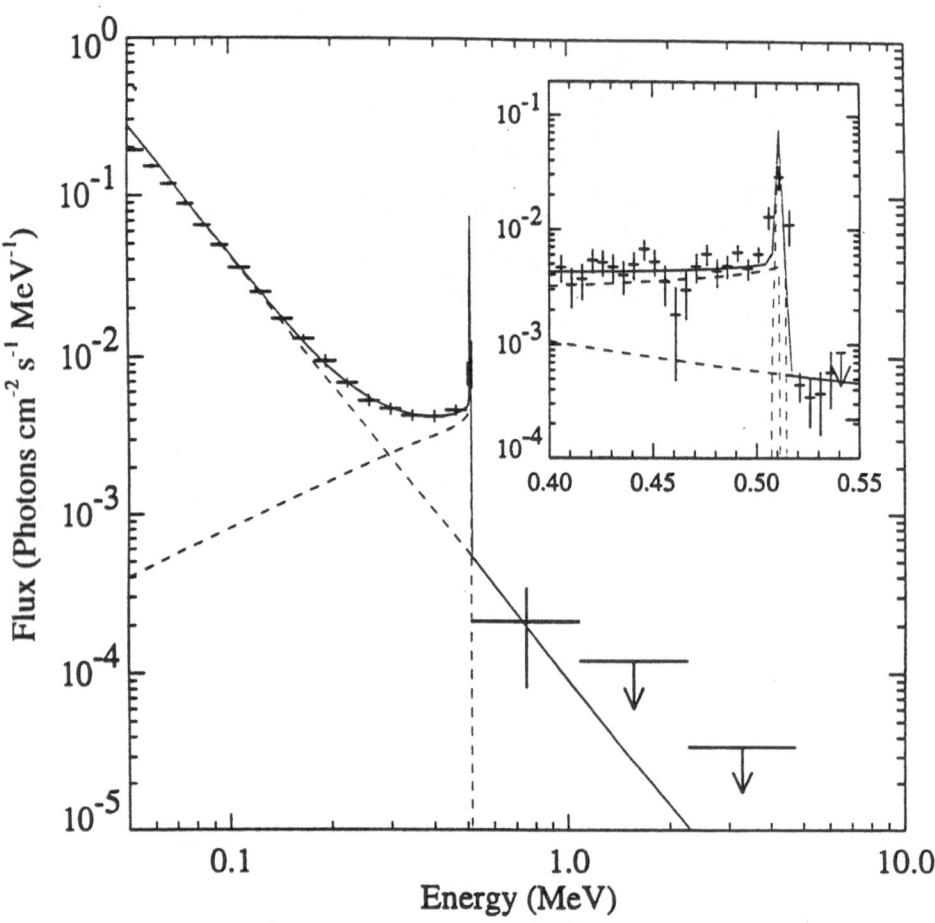

Figure 2–2: The OSSE Galactic center spectrum summed from all four detectors over the period 1991 July 13–24. The fitted lines represent a power law, a photopeak line and a positronium continuum.

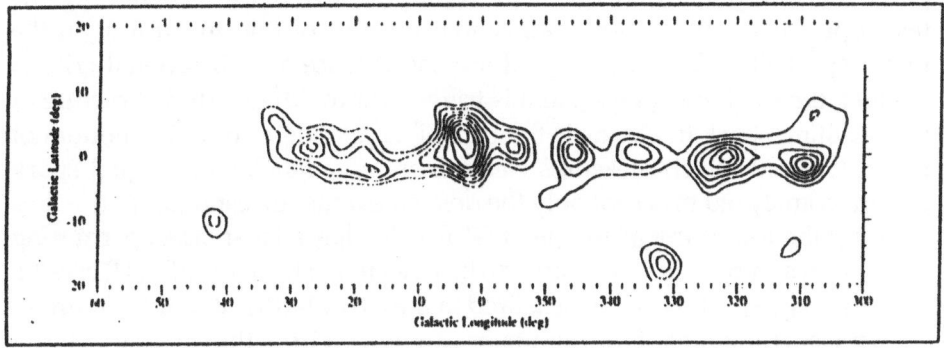

Figure 2–3: Intensity map for 1.8 MeV emission along the Galactic plane, derived from the combined plane observations of the CGRO sky Survey. Contour levels are in steps of 2×10^{-4} cm^{-2} sr^{-1} s^{-1}.

absorbed in one of fourteen (high-Z) NaI(Tl) scintillators. The unusual aspect of this detector is that the location of the γ-ray on the sky is given by a annulus and not a point. The correlation of many events enables the localization of point sources or the creation of sky maps. Time-of-flight measurements, pulse shape discrimination and anticoincidence shields are used to reject background events. CompTel has a wide field of view (about 1 sr) and an angular resolution under optimal conditions of about 1°. Its energy resolution is about 9% at 1 MeV. During Phase 1 of the mission, CompTel completed the first all–sky survey in the energy range of 1–30 MeV. In addition to its *Compton* telescope mode capabilities, CompTel can measure energy spectra of solar flares or bright cosmic γ-ray bursts between 0.1 and 10 MeV in two of its lower detector plane modules, and it can measure neutrons from solar flares in a special telescope mode. Figure 2–3 is an intensity map of the Galactic plane for 1.8 MeV line emission. The Principal Investigator of CompTel is Dr. V. Schönfelder of the Max-Planck Institute for Extraterrestrial Physics.

2.3   EGRET

The Energetic Gamma Ray Experiment Telescope (EGRET) is the highest energy instrument on *Compton* Observatory , and covers the broadest energy range, from 20 MeV to 30 GeV. It has a wide field-of-view (about 0.6 sr), and can localize point sources to an accuracy of 5–30 minutes of arc. Background rejection is highly efficient. The energy resolution is about 20% in the 100–200 MeV range. Pair production is the dominant photon-matter interaction process at the photon energies covered by EGRET. A γ-ray entering the telescope creates an $e^- - e^+$ pair within a series of thin metal foils in the upper spark chamber assembly. If anticoincidence and directional criteria are met, the track imaging system is triggered providing a digital picture of the resulting spark tracks, as is the NaI(Tl) crystal detector at the bottom of the instrument which measures electron energy. Finally, the digital spark picture is analyzed to accept only the desired events corresponding to γ-rays entering the top of the telescope. EGRET also has a *burst mode*, permitting solar flares and cosmic γ-ray bursts to be detected in the large NaI(Tl) crystal, where energy spectra can be measured in the 0.6 to 140 MeV range. Figure 2–4 is a phase histogram of the high-energy emission from the Geminga pulsar. The Co-Principal Investigators of EGRET are Dr. C. E. Fichtel of the NASA Goddard Space Flight Center and Dr. K Pinkau of the Max-Planck Institute for Plasma Physics.

2.4   BATSE

The Burst and Transient Source Experiment (BATSE) consists of eight identical modules located on each corner of the spacecraft. This configuration provides continuous viewing of nearly the entire sky. BATSE is optimized to measure brightness variations on microsecond timescales of 0.03 to 1.9 MeV γ-rays. Each module contains a Large Area Detector, containing a 50 cm diameter by 1.3 cm thick NaI(Tl) crystal, and a Spectroscopy Detector, containing a 13 cm diameter by 8 cm thick NaI(Tl) crystal. The Spectroscopy Detector is sensitive to 0.015 to 110 MeV γ-rays with a spectral resolution of 6%. The BATSE electronics identifies γ-ray bursts and solar flares and then collects and stores the data in various formats in *burst memory* for later telemetry. Additionally, continuous data, sampled at ≈ 1 s intervals, and on-board epoch-folded data can be obtained to study pulsed sources, long-lived transients and steady sources. When a γ-ray burst is identified by BATSE, a burst-trigger is sent to the other experiments, allowing them to switch to a burst mode for data collection. For solar flares, a solar-flare trigger is sent to OSSE and CompTel. OSSE may respond by slewing towards the sun. Figure 2–5 shows the distribution in Galactic coordinates of 687 γ-ray bursts. The BATSE Principal Investigator is Dr. G. J. Fishman of the NASA Marshall

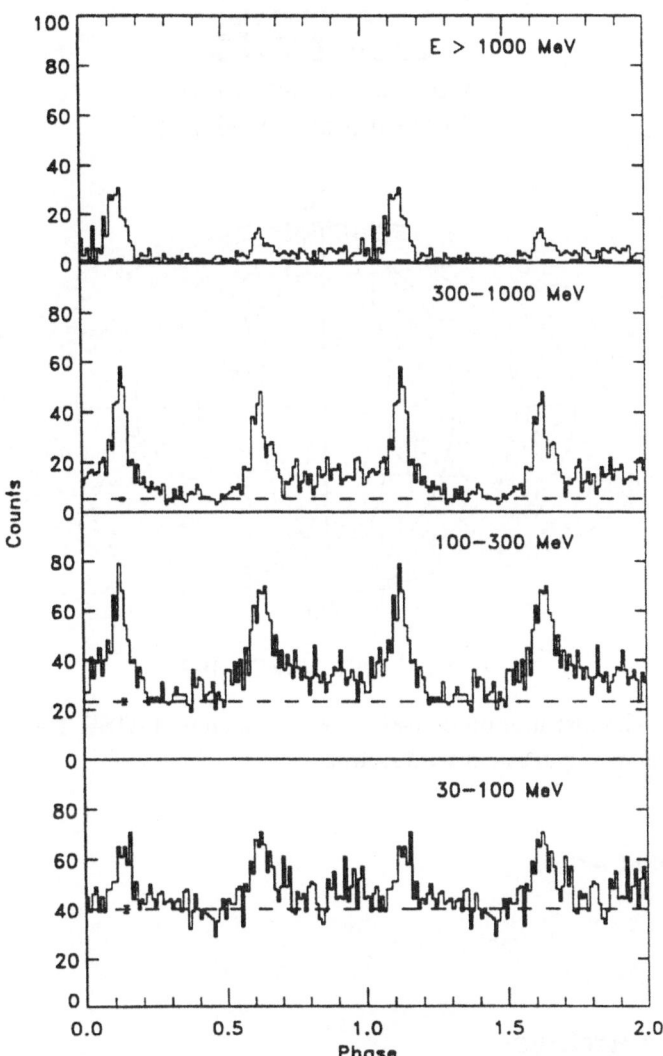

Figure 2–4: The phase histogram of the Geminga pulsar in four energy ranges. All EGRET observations from phase 1 and 2 were used.

BATSE
Compton Observatory
1000 Gamma-Ray Bursts

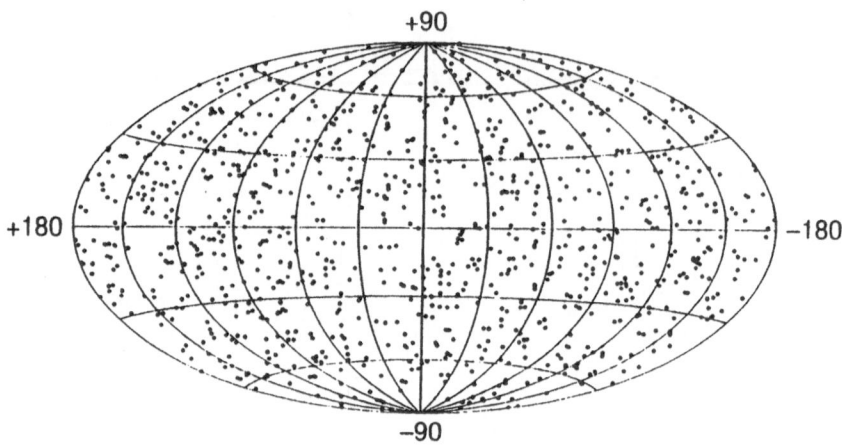

April 21, 1991–May 27, 1994

Figure 2–5: The distribution of γ-ray bursts as seen by BATSE. The angular distribution of bursts is consistent with isotropy.

Space Flight Center.

2–3 The Archive

The COSSC archive contains public-domain data products from each of the four instrument teams. Those data products held by the COSSC as of April 1994 are listed in Table 2–2. As a rule-of-thumb the data is generally released into the public domain about 15 months after the observation is made, but the release date may slip due to the complex nature of the instruments and the data processing. Almost all of the phase 1 survey data is currently available.

The type of data products range from low-level photon event lists to high-level sky maps. All data are store as compressed FITS files as mandated by NASA. A description and typical uses of each data product are given in the following sections. For a brief description of each, see table 2–2.

3.1 CONTENTS

3.1.1 BATSE IBDB The BATSE Individual Burst Database contains data of those BATSE triggers (of bursts) determined by the BATSE standard processing software to be solar flares or γ-ray bursts. The standard processing software produces a file set of about 30 files for each burst. The file set contains data from 4 of the 8 Large Area Detectors (LADs) and Spectroscopic Detectors (SDs) as well as several types of discriminator data and time-tagged event data. This data is almost exclusively used for the analysis of solar flares and γ-ray bursts.

3.1.2 BATSE CONT The BATSE Continuous Database contains High Energy Resolution (HER) and Spectroscopic High Energy Resolution (SHER) data from each of the 8 BATSE detectors for a total of 16 files per file set. The 8 detectors provide continuous coverage at a time resolution of 1.024 s of the entire sky, less Earth blockage. This database is proving to be both valuable and versatile. For example, the data is being used *i)* to discover and locate hard X-ray pulsars whose spin period is greater than 1 s, *ii)* to discover and locate hard X-ray sources and transients by use of Earth occultation techniques, *iii)* to monitor known hard X-ray transients on a daily basis, and *iv)* to study the hard X-ray emission from the sun on a daily basis.

3.1.3 CompTel EVP The CompTel Photon Event List Database contains a list of all events that were accepted by the CompTel standard processing software during a viewing period. This is a low-level data product and the files are generally quite large, ≈60 MB. Each record contains the event's time, direction, scattering angle, and energy. These files can be used for detailed timing, imaging, and spectral analyses, though such analyses probably require special software, due to the unique nature of this instrument.

3.1.4 CompTel DRE The CompTel Binned Event Matrix is a higher-level product based on the CompTel Photon Event List. Each file set contains four files, one for each standard energy range, *i.e.* 0.75–1.0 MeV, 1–3 MeV, 3–10 MeV, and 10–30 MeV. These files are useful for image studies and source detection.

3.1.5 EGRET SDB The EGRET Summary Database is a high-level data product containing the photon event list for a viewing period. Each record includes the events time, energy, and position, either in equatorial and galactic coordinates. These files are useful for timing, imaging, and spectral analyses, for example, studies of pulsar timings or γ-ray transients.

3.1.6 EGRET EXP The EGRET Exposure History Database is a low-level data product. It chronicles the on-and-off modes of the nine subtelescopes.

This data can be used to track the exposure history of a point source, ie. can be used to estimate the amount of observing time a source has received. To create an exposure map, the effective area and sensitivity calibration files are also needed.

3.1.7 EGRET MAP The EGRET Sky Maps Database is a high-level data product containing count, exposure, and intensity maps in 4 energy ranges for that part of the sky observed during a viewing period. These maps are created by the EGRET standard processing software using SDB and EXP databases and calibration files. This data is useful for source detection and for determining source intensities or upper-limits.

3.1.8 OSSE SDB The OSSE Spectral Database is a low-level data product providing data from all 4 OSSE detectors on a daily basis. The file set is quite large ($\approx$20 MB). Use of this data is limited, due to the complexity of background subtraction and source confusion, and analysis of the data is probably best done using the OSSE analysis software.

2–4   Future Plans

The data products discussed in this chapter should be considered only the beginning of the *Compton* Observatory Archive. The archive presently contains $\approx$10 GB of uncompressed data. A rough estimate of the archive's volume from the first six years of the mission is of the order of a few 100 GB. As the instrument teams make available to the public other data products, these too will be added to the archive.

The COSSC is also working to make available analysis software for each instrument, either by providing access to the instrument team's computers or by installing the instrument teams's software on COSSC's computers. Access to the instrument teams's analysis software at the COSSC is via GammaCore which provides a graphical interface to access the available γ-ray analysis software. The user can use the mouse to start an application or display a its list of options.

Access Pointers

[→1] Compton Observatory Science Support Center:
http://cossc.gsfc.nasa.gov/cossc/cossc.html

Table 2–1: Summary of Compton Observatory Instruments

	OSSE	COMPTEL	EGRET	BATSE LAD[1]	BATSE SD[2]
Energy Range (MeV)	0.05–10	0.8–30	20–30000	0.02–1.9	0.01–100
Effective Area (cm²)	550 (0.2 MeV)[3] 360 (1.0 MeV)[3] 140 (5.0 MeV)[3]	25.8 (1.27 MeV) 29.3 (2.75 MeV) 29.4 (4.43 MeV)	1100 (100 MeV) 1600 (500 MeV) 1100 (3000 MeV)	1000 (0.03 MeV)[3] 1800 (0.10 MeV)[3] 550 (0.66 MeV)[3]	100 (0.3 MeV)[3] 127 (0.2 MeV)[3] 52 (3.0 MeV)[3]
Field-of-View (sr)	0.013 (3.8° × 11.4°)	≈ 1.1	≈ 0.6	4π	4π
Effective Geometry (cm² sr)	13	30	1050 (≈500 MeV)	15000	5000
Source Localization[6] (arc-min)	10 (scanning)	60	10	180 (strong burst)	
Line Sensitivity[5] ($\times 10^{-6}\ cm^{-2}\ s^{-1}$)	65 (0.2 MeV) 90 (1.0 MeV) 40 (5.0 MeV)	60 (1.0 MeV) 15 (7.0 MeV)			0.4% EW (5 s integration)
Continuum Sensitivity[5] ($\times 10^{-6}\ cm^{-2}\ s^{-1}$)	650 (0.2 MeV) 350 (1.0 MeV) 70 (5.0 MeV)	170 (2.0 MeV) 55 (5.0 MeV) 9.4 (20 MeV)	0.13 (100 MeV) 0.02 (500 MeV) 0.01 (3000 MeV)	0.03 (1 s burst)	
Spectral Resolution (FWHM)	12.5% (0.2 MeV) 6.8% (1.0 MeV) 4.0% (5.0 MeV)	8.8% (1.27 MeV) 6.5% (2.75 MeV) 6.3% (4.43 MeV)	~20%	32% (0.06 MeV) 27% (0.09 MeV) 20% (0.66 MeV)	8.2% (0.09 MeV) 7.2% (0.66 MeV) 5.8% (1.17 MeV)
Temporal Resolution[10] (ms)	2048 (normal) 125 (pulsar) 4 (burst)	125	125	1024 (DISCLA[7]) 0.001 (TTE) 0.002 (burst)	2048 (DISCSP[8]) 0.128 (burst)

Notes: [1] LAD = Large Area Detector, [2] SD = Spectroscopy Detector, [3] per detector, [4] for a strong point source, [5] for a 5 × 10⁵ s observation, off the Galactic Plane, [6] for a 0.1 ×Crab source and 1 σ error, [7] DISCLA = LAD Discriminator data, [8] DISCSP = SD Discriminator data, [9] TTE = Time-Tagged Event data, [10] accurate to 125 ms with respect to UTC.

Table 2–2: Summary of COSSC Archive Data Products

Instrument	Product Name	Product Level	Files per File Set	Viewing Periods	Description
OSSE	SDB	low	6	0.2-201	The **Spectral Database** contains 2 minute spectra, ^{60}Co spectra, burst data, and standard plot files from the 4 detectors, grouped on a daily basis.
COMPTEL	EVP	low	1	1-13.5	Time tagged photon event lists for a given viewing period. These files are very large ($\sim$ 60 MB) and intended only for detailed low-level timing, spectral and imaging studies.
COMPTEL	DRE	high	4	1-4	Binned event matrices for a given CGRO viewing period. These files come in 4 standard energy ranges (0.75-1.0 MeV, 1-3 MeV, 3-10 MeV, 10-30 MeV).
EGRET	SDB	low	1	0.2-206	The **Summary Database** contains time tagged photon event lists for a given viewing period.
EGRET	MAP	high	12	0.2-206	The **Sky maps** are count, exposure and intensity maps in 4 different binning schemes for the region of sky covered by a given viewing period.
EGRET	EXP	low	1	0.2-206	The **Exposure History** has information necessary to compute EGRET exposures for a given viewing period.
BATSE	IBDB	low	30	0.2-31	The **Individual Burst Database** includes time of trigger and derived source location, and all associated burst mode data (DISCSC[1], PREB[2], MER[3], TTS[4], TTE[5], STTE[6], HERB[7], SHERB[8] as collected) filtered for data quality, for a given γ-ray burst or solar flare. Relevant background data (DISCLA[9], DISCSP[10], CONT[11], HER[12], SHER[13]), associated spacecraft position and attitude information is included with each of these products.
BATSE	CONT	low	16	0.2-31	The **Continuous Data** consists of the daily HER[12] and SHER[13] from the 8 BATSE detectors with a time resolution of 1024 ms.

Notes: [1]DISCSC = discriminator science data, [2]PREB = pre-burst data, [3]MER = Medium Energy Resolution data, [4]TTS = Time-to-Spill data, [5]TTE = time-tagged event data, [6]STTE = spectroscopic time-tagged event data, [7]HERB = high energy resolution burst data, [8]SHERB = spectroscopic high energy resolution burst data, [9]DISCLA = Large Area Detector discriminator data, [10]CONT = continuous data, [11]DISCSP = Spectroscopy Detector discriminator data, [12]HER = high energy resolution data, [13]SHER = spectroscopic high energy resolution data.

3

Data from the X-ray Observatory ROSAT

H. U. Zimmermann, H. Brunner, R. Pisarski, W. Voges, and
M. G. Watson

3–1   The ROSAT Mission

1.1   INTRODUCTION

On June 1st, 1990, the ROSAT observatory was launched successfully into
a circular orbit of 575 km altitude and 53 degrees inclination. The satellite
carries two telescope systems on-board: an X-ray telescope (XRT), operating
in the energy range 0.1 to 2.4 keV, and a smaller XUV telescope (the ROSAT
Wide Field Camera, WFC) sensitive in the range from about 20 to 200 eV.

ROSAT —an acronym for Röntgen Satellite— is a German project with
major contributions from the United States and the United Kingdom (Trümper
1983). The scientific mission responsibility lies with the Max-Planck-Institut
für Extraterrestrische Physik (MPE) at Garching, Germany. Germany (DARA,
MPE) built the satellite and the X-ray telescope including 2 of the 3 focal
plane detectors. The US (NASA, SAO) provided one focal plane detector
for the X-ray telescope and launched the satellite. The WFC was built by
a consortium of five British institutes (SERC, the Universities of Leicester
and Birmingham, the Mullard Space Science Laboratory, Imperial College of
Science Technology and Medicine and the Rutherford Appleton Laboratory).

On July 30th, 1990, ROSAT started its main scientific task: to perform
the first all-sky survey with an imaging telescope both in the soft X-ray and
the XUV regimes. After this survey, which lasted 6 months, an extended
observatory program on a guest observer basis was begun with pointings on
selected targets.

In 4 years of operation, ROSAT detected on the order of 100,000 X-ray
sources (before ROSAT $\sim$ 5000 were known), comprising all object types
from normal stars up to distant quasars. Besides the $\sim$ 60,000 sources seen
in the all-sky survey, where mean exposure times of the order 500 sec were
achieved, many more serendipitous sources were found in the $\sim$ 5000 pointings to special targets where mean exposures are $\geq$ 8000 sec. Thus an enor-

D. Egret and M. A. Albrecht (eds.), Information & On-Line Data in Astronomy, 27–35.

mous X-ray data base is available for detailed studies of specific objects and different object classes.

With the WFC instrument about 500 new EUV sources were detected (before ROSAT only a handful of EUV sources were known), mainly during the all-sky survey when the detector had still its full sensitivity.

Important results also come from mapping the diffuse X-ray and EUV backgrounds with high sensitivity and high spatial resolution.

1.2 THE INSTRUMENTS

The X-ray telescope (XRT) consists of a fourfold nested Wolter I type mirror (Aschenbach 1988) with an entrance diameter of 80 centimeters and 3 focal detectors. Two of these are position-sensitive proportional counters (PSPC) (Pfeffermann et al. 1986) with a positional resolution of about 20 arcsec and a moderate energy resolution resulting in 4 to 5 spectral bands within the 0.1 to 2.4 keV energy range. A boron filter can be additionally used to enhance the resolution at lower energies. This detector has a field of view of 2 degrees and has been used during the all-sky survey and for more than 80% of the pointed observations. The third detector is the High Resolution Imager (HRI), a channel plate detector allowing spatial resolution of a few arcsec, but providing almost no energy information. It is a rebuild of the HRI on-board the *Einstein* observatory. The Guiness Book of Records lists the XRT as the largest and most accurate X-ray telescope built to date.

The Wide Field Camera (WFC) is a threefold nested Wolter-Schwarzschild I telescope with 2 channel plate detectors in the focal plane (Barstow et al., 1988). The positional resolution is about 1 arcmin in the centre of the 5 degree-wide field of view. Filters defining 4 wavelength bands provide spectral information.

1.3 THE ROSAT SCIENCE DATA CENTRES (RSDC)

Located at MPE, the German RSDC is the central interface between the observatory and the other ROSAT Data Centres in the UK (RAL/Leicester), the USA (GSFC/SAO) and the German WFC centre (AIT). All data centres serve as direct contact points for the ROSAT user community in their countries.

Each of the data centres fulfills its special tasks within the project. For example the main activities of the German RSDC are briefly outlined here (Zimmermann et al., 1986):

> ▷ The centre provides organisational, editorial and technical support during the *call for proposals* and the *proposal selection phases*. It has full responsibility for the mission timeline production and the mission scheduling process.

▷ XRT instrument control and detailed checkout of all relevant XRT data are performed in real and near real time.

▷ The RSDC has full responsibility for the development and implementation of the Standard Analysis Software System (SASS) used at both the German and the US RSDC for the routine processing of all XRT data. The standard data processing guarantees a uniform calibration and provides observers with the results of a standard image and source analysis of each pointed observation.

▷ Sophisticated analysis software (EXSAS) is maintained and distributed by the centre to support users in the evaluation of ROSAT data at their home institutions or at the working facilities in the RSDC.

▷ The centre provides extended observer support, offering advice, documentation and other information services as well as maintaining the ROSAT Archives.

1.4 Instrument and Mission Status

Originally planned as a mission of 20 months, ROSAT has performed extremely successfully for more than 4 years to date. Now the gas resources for the PSPC detector are exhausted and future observations will therefore be done with the HRI detector only. Failures of a few satellite components (especially in the attitude control system) caused some restrictions with observation scheduling, but these were compensated almost entirely by sophisticated reallocation of on-board resources (sensors, computer programs). Only the sensitivity of the WFC detectors was affected more seriously at the end of January 1991.

3–2 ROSAT Data

2.1 Data Flow

The detectors in the focal planes of the X-ray and EUV mirrors record each photon arriving in the detector separately. The PSPC instrument, for example, delivers a 2-dimensional position of each photon event in detector coordinates, the pulse-height of the event as a measure of its energy, and the arrival time. The photon event data are immediately written to the on-board tape recorders together with housekeeping and attitude information. During the 5 to 6 consecutive contacts per day between the satellite and its ground station —the German Space Operation Center (GSOC) at Oberpfaffenhofen/Weilheim— the on-board recorded data are transmitted to the ground. About 20% of the data are immediately passed on via a dedicated line to the German RSDC for instrument health and data quality

checks. A few days later the full set of telemetry data plus additional auxiliary information on orbit, attitude, commands and the mission program is
delivered to the RSDCs in Germany, the UK and the USA.

2.2   STANDARD DATA PROCESSING

For all X-ray data a Standard Data Processing (Standard Analysis Software
System SASS, developed by MPE and SAO; Voges *et al.* 1992) is performed
in both the German and the US RSDC. During this process the basic event
information on each photon first undergoes various corrections and normal-
izations (e.g. detector linearisation and pulse-height calibration). The photon
event files containing this information form, together with housekeeping and
attitude data, the basic input to all further analysis.

As a next step in the Standard Processing, images of the observed sky
regions are formed. Different techniques (sliding window and maximum
likelihood methods) are used to search the images for point-like sources
above a specified significance level. Best estimates for the position of each
detected source, the size of the corresponding error box, the intensities in
different energy bands and a source extension indicator are the output of the
final maximum likelihood algorithm. For stronger sources, in addition, a
search for time variability is performed and spectra are fitted with standard
emission models. Finally, a large optical catalogue (SIMBAD) is scanned for
possible counterparts of the detected X-ray objects. An extended protocol
of the results from the Standard Processing is produced and distributed —
together with the standard data sets — to the principal investigators of the
observations. Thereafter all data sets enter the internal ROSAT archive.

In order to guarantee computer system independence for the data distribution, all data sent to the observers are delivered in FITS format (Flexible
Image Transport System: a data transfer standard widely accepted by the
astronomical community). The data sets for each observation consist of different tables and images containing all the basic and auxiliary data needed
for further detailed analysis: primary photon event data, information on
instrument housekeeping and quality parameters, attitude and orbit files,
calibration data and also selected results from the standard analysis performed at the RSDC.

Processing of WFC data carried out at the WFC project centres in the UK,
and also at the German WFC data centre, is similar to that described for the
XRT data.

2.3   INTERACTIVE ANALYSIS TOOLS

It has been widely recognized in the past that archives containing raw or pre-
processed data can be used effectively only if appropriate analysis software

is commonly available. For ROSAT data major interactive analysis systems have been developed.

The Extended Scientific Analysis System (EXSAS) developed by the German RSDC (Zimmermann et al. 1992), is a general software package for interactive analysis of X-ray data, with particular emphasis on the ROSAT XRT and WFC instruments. This large collection of application software modules is embedded in the astronomical image processing system MIDAS (ESO). EXSAS/MIDAS is highly portable and has been installed on both VMS and different UNIX operating systems. EXSAS software may be requested from the German RSDC.

The Post Reduction Offline System (PROS) is the US equivalent to EXSAS, developed by SAO at Cambridge (USA) (Worrall et al. 1994). It runs in the IRAF environment (Kitt Peak Observatory) and is available from SAO.

In the UK the ASTERIX system, developed by the UK ROSAT project, provides similar functionality as EXSAS or PROS within the UK's Starlink software environment.

3–3   ROSAT On-Line Services

Both the German RSDC (MPE) and the US RSDC (together with HEASARC, the High Energy Astrophysics Science Archive Research Center, at GSFC) maintain extended information and data services for all major mission aspects. Similar services are provided in the UK by the University of Leicester which has responsibility for support of UK guest observers and also runs the UK ROSAT Data Archive Centre (UKRDAC). The German WFC Data Centre (AIT) also offers WFC related information and on-line data services.

3.1   MAJOR SERVICES AVAILABLE

All centres continuously enhance their services which differ slightly from centre to centre. Typical installations are listed below:

- ▷ general e-mail service
- ▷ general ROSAT info (status, news, publication lists)
- ▷ proposal support (utilities, documentation, advice)
- ▷ ROSAT mission timelines, observation logs
- ▷ instrument calibration issues
- ▷ Standard Processing status, data shipping lists, bug reports
- ▷ interactive analysis software distribution, updates, bug reports
- ▷ public archive related information and utilities

▷ ROSAT Data Archive on-line (and off-line) services

▷ distribution of project news via electronic mail

▷ project documentation

3.2   ACCESS MODES

There are several possibilities of directly connecting to the on-line services:

▷ The most general one is via anonymous ftp. This access mode allows to read and copy available information including data sets from the ROSAT Data Archive.

▷ Access to some information services via World-Wide Web (WWW) services like NCSA Mosaic has recently been enabled. The possibilities are rapidly evolving.

▷ Some specific services are also accessible via the BROWSE (originally developed as the EXOSAT database system) and the Mission Information and Planning System (MIPS) —developed by GSFC— utilities, installed at several centres.

▷ Some RSDCs provide additional local access utilities.

Details on how the major services can be accessed are given in the technical annex.

3–4   ROSAT Archives

4.1   DATA RIGHTS

The data rights of the ROSAT X-ray all-sky survey are with MPE. Scientists at MPE collaborate with many groups outside the institute on the analysis of this large database. A first source catalogue is in preparation and will then be accessible via the ROSAT Archives. Similarly, the WFC data from the all-sky survey belong exclusively to the WFC consortium institutes. A first WFC source catalogue has already been published (Pounds et al. 1993). It is planned that the WFC survey data will become public in the near future.

For the pointed observation phases, all observations originate from proposals sent to one of the data centres in Germany, the UK or the US and screened by both national and international selection committees. Data from the observations, as soon as they become available, are then distributed automatically to the corresponding principal investigator (PI). Usually these data belong exclusively to the PI for one year. In cases, where an observation with priority A or B has achieved only less than 70 % of the scheduled on-target time, the one year proprietory period begins only after automatic follow-up observations bring it to at least the 70 % limit. Due to these rules, data will not enter the ROSAT Archives before the above mentioned rights on the data have expired.

In spring 1994 a reprocessing of all data from pointed observations started. This corrects some early processing errors. For the affected datasets, for which the initial processing (REF 0 processing) took place before January 1993, the reprocessed version will belong to the PI exclusively for a half year period before entering the public archives.

4.2 INTERNAL ROSAT ARCHIVE (XRT)

Primary, intermediate and final products of the Standard Data Processing (about 0.5 GBytes/day) enter an internal archive at the RSDCs. This archive is for internal usage only and serves as source for the Data and Result Archives.

4.3 ROSAT DATA ARCHIVE FOR XRT OBSERVATIONS

Since the end of 1992 a public ROSAT Data Archive for the X-ray data is maintained at both the German RSDC (MPE) and the US RSDC (GSFC) in identical form. An additional copy of the Archive is installed at the UKRDAC (UL).

The ROSAT Data Archive contains processed data sets of all pointed ROSAT XRT observations for which the original data rights of the principal investigator have expired. The data are a subset of those initially delivered to the PI and comprise primary and secondary data products necessary for detailed data analysis. Not included are the result files of the Standard Analysis Processing, since the initial processing was not regarded as sufficiently reliable for general unscreened usage (see ROSAT Result Archive).

Data sets are all in FITS format. The initial FITS implementations in Germany and the USA were different. From the beginning of the reprocessing of all pointed observations, data sets are now delivered in a new common FITS format, the Rationalized Data Format (RDF) developed by GSFC. Until the reprocessing is complete — expected in the second half of 1995 — the archive user will therefore find ROSAT XRT data in 3 different FITS structures and with slightly varying contents. This is usually not a problem because the analysis systems of the RSDCs can handle all 3 formats and the most important data sets (photon event, attitude, orbit and housekeeping files) are always present:

    ▷ the FITS structure, as formerly developed by the German RSDC for the use with EXSAS, provides up to 20 different data files for every PSPC observation and 12 for each HRI observation.

    ▷ the US RSDC released data originally in a FITS form developed for the

PROS analysis system. It consists of up to 20 files for the PSPC and 26 files for the HRI.

▷ the new RDF format consists of a few main FITS files that expand on input to a multitude of individual files. It provides the most complete set of basic and auxiliary data files for an XRT observation.

4.4   DATA ARCHIVE FOR WFC OBSERVATIONS

A similar data archive for the pointed WFC data has been installed in the UK and in Germany. The installation at the UKRDAC provides the ROSAT data in the Starlink HDS format, while the German data centre at Tübingen has data formatted in the EXSAS compatible FITS format as originally used by the German RSDC. The UKRDAC plans to convert all the WFC archive data to FITS format (compliant with the RDF) over the next year. Images from pointed WFC observations are already available in RDF FITS versions.

4.5   ROSAT RESULT ARCHIVE FOR XRT OBSERVATIONS

The ROSAT Result Archive (RRA) is expected to open towards the end of 1994. It will contain selected results and data sets taken from the Standard Processing, both for the observed fields and the individual sources detected therein. To ensure a homogeneous quality throughout the archive, teams of experts will thoroughly inspect all data entered. This is necessary, because the Standard Processing, although quite sophisticated in its analysis, is a fully automated process that may have difficulties in cases of unusual data, e.g. in extended emission regions.

The database will offer a unique possibility to access the major results of all ROSAT pointings in an easy way. The best way to access the database will be through the 'browse' utility (EXOSAT Database or HEASARC system), installed at GSFC, MPE and at Leicester. It provides an astronomer-friendly environment to retrieve and select X-ray and other astronomical data. Included are utilities to compare and, to some extent even analyze retrieved data.

The community will be notified as soon as the Result Archive will become available.

4.6   POSSIBLE PUBLICATION CONFLICTS

Every user of the ROSAT Data Archives is strongly urged to find out the status of the work carried out by the original principal investigator on the relevant data. This will avoid duplication of efforts in the analysis and possible publication conflicts.

References

[1] Aschenbach, B., 1988, Appl. Optics, vol. 27, pp. 1404-1413

[2] Barstow, M., and Willingale, R., 1988, JBIS, Vol. 41, p. 345

[3] Pfeffermann, E., Briel, U.G., Hippmann, H., Kettenring, G., Metzner, G., Predehl, P., Reger, G., Stephan, K.-H., Zombeck, M.V., Chappell, J., and Murray, S.S., 1986, SPIE Vol.733 Soft X-Ray Optics and Technology, p. 519

[4] Pounds K.A. et al. 1993, MNRAS 260,77

[5] Trümper, J., 1983, Adv. Space Res. vol.2, No. 4, pp. 241-249

[6] Voges, W., Gruber, R., Paul, J., Bickert, K., Bohnet, A., Bursik, J., Dennerl, K., Englhauser, J., Hartner, G., Jennert, W., Koehler, H., Rosso, C., 1992, in Proceedings on 'European ISY meeting: Symposium', Munich

[7] Worrall, D. M., et al., 1994, to appear in Data Analysis in Astronomy IV

[8] Zimmermann, H.-U., Gruber, R., Hasinger, G., Paul, J., Schmitt, J., and Voges, W., 1986, Data Analysis in Astronomy II, p. 155

[9] Zimmermann, H.-U., Belloni, T., Boese, G., Izzo, C., Kahabka, P., Schwentker, O., 1992, in Data Analysis in Astronomy IV, Plenum Press, N.Y., ed. by DiGesu, V., Scarsi, L., Buccheri, R., Crane, P., Maccarone, M. C., and Zimmermann, H.-U., p.141

Access Pointers

[→1] German ROSAT Scientific Data Center (MPE Garching):
     http://rosat_svc.mpe-garching.mpg.de/

[→2] GSFC/NASA ROSAT Scientific Data Center (HEASARC):
     http://heasarc.gsfc.nasa.gov/ and MIPS:
     http://acadia.gsfc.nasa.gov/adf/rosatDC.html

[→3] UK ROSAT Data Archive Centre (Tübingen):
     http://www.star.le.ac.uk/

4

The Einstein On-Line Service

Daniel E. Harris & Carolyn Stern Grant

4-1 General

In January 1989, the High Energy Astrophysics Division of the Smithsonian Astrophysical Observatory established an on-line service to help astronomers prepare ROSAT proposals by providing access to the preliminary source list from the IPC (Imaging Proportional Counter) catalog project of the Einstein Observatory. Since then, we have added many more source tables (including ground based data), and provided access to Einstein images for downloading. The user interface provides easy access to extensive documentation and several options for search and retrieval.

Although we have improved the functionality and made significant additions to the databases, we still maintain a simple menu interface accessible from any type of terminal. We do not provide batch processing of queries because outside files are not permitted in the restricted EOLS environment (*i.e.* no uploading).

EOLS provides all of our databases to the NASA Astrophysics Data System (ADS) and our documents which describe each table are thus written in the ADS format. In conjunction with theIAU working group on Archiving and Data Bases for Radio Astronomy, EOLS serves as an experimental platform for on-line access to the radio source lists collected by H. Andernach.

4-2 Services and contents

2.1 THE DATABASES

Initially, the source tables were mostly related to those from the Einstein Observatory. However, since it is relatively easy to ingest data once the required software has been written, we have been able to expand our searchable tables significantly. The main requirement is a reliable document describing the data.

D. Egret and M. A. Albrecht (eds.), Information & On-Line Data in Astronomy, 37–41.
© 1995 *Kluwer Academic Publishers.*

Our database tables are divided into "affinity groups" for ease of assimilation. Currently we have 128 tables with over 700,000 entries, divided into the following topics:

2Ecat	Einstein Observatory Catalog of IPC X-ray Sources
Esurveys	Tables of IPC and HRI surveys (Deep, Medium)
Galaxies	Einstein Galaxy Lists
Hri	HRI Field Centers and Source List
Ipc	Mostly Rev 1B (results of standard processing)
Ocat	Einstein Observatory Pointing Catalog ("Yellow Book")
Quasars	Einstein and Other Quasar Lists
Radio	Radio Source Catalogs
Rosat	ROSAT Target List and Timeline
Slew	Slew Survey Source List
Snr	Supernova Remnants (Einstein)
Stellar	Assorted Einstein Observatory Results on Stars

2.2   DOCUMENTATION

There is a document describing each database table and there are many others with extensive detail such as Vol. 1 of the "Einstein Observatory Catalog of IPC X-ray Sources". The subdirectory structure mimics that of the database affinity groups, with additional subdirectories for archival documents (e.g. the Revised User's Manual, HEAO Newsletters, etc.) and the help documents in the "How-to" subdirectory.

2.3   DATA PRODUCTS

All of our Einstein CD-ROMs are permanently mounted for downloading images. They may be obtained via anonymous ftp or with the VMS COPY command. Thus when an occasional image is required, it can be obtained at any time without the need of a cdrom reader and the appropriate cdrom. The 2E catalog cdroms contain primary arrays of smoothed IPC fields that have been corrected for background and vignetting. The HRI versions are primary FITS arrays of photons. The more recent additions for both the IPC and HRI give the complete data in the form of binary extension tables of photons; hence these are suitable for spectral and temporal analyses. As of April 1, 1994, there are five sets, listed in Table 4–1.

2.4   COMMUNICATION

This section of the EOLS features a bulletin board with news on ROSAT, other missions, and updates to Einstein material. An e-mail facility is provided for sending documents and the results of database searches to yourself or

Title	Data Set Type	Date
The 2E Catalog of IPC X-Ray Sources	FITS smoothed arrays	1 Jun 1991
The Database of HRI X-Ray Images	FITS photon arrays	1 Jun 1990
The IPC Slew Survey	FITS binary tables	1 Jan 1991
The HRI Images in Event List Format	FITS binary tables	1 Jan 1992
The IPC Images in Event List Format	FITS binary tables	1 Jun 1992

Table 4–1: CD-ROMs Mounted on EOLS

others.

2.5 TOOLS

Current data tools include a precession routine and a method of interpolating HI column densities of galactic HI for declinations north of -40 degrees (the Bell labs survey limit).

4–3 Description of operation

3.1 ACCESS TO THE EINSTEIN ON-LINE SERVICE

To reach the EOLS, we provide several methods:

MODEM (300/1200/2400 BAUD): dial +1 (617) 495-7047 or 495-7048. Our test used odd parity, stop bits=1, and data bits = 7.

TELNET: the address is einline.harvard.edu, alias cfa204.harvard.edu, the current node number is 128.103.40.204, but this may change if the local network is reconfigured.

SPAN: set host 6714, or if your name server is up to date set host cfa204

Once successful, the login name is "einline"; no password. NB: Since UNIX is case sensitive, make sure you use lower case when logging in!

WORLD-WIDE-WEB: The einline home page contains a link which will spawn an interactive session with EOLS [→1]. The url is:

http://hea-www.harvard.edu/einline/einline.html

3.2 DATABASE QUERIES

To minimize the nuisance of building queries, we have provided "canned" queries with fixed retrieval columns for each database ("Quick Queries", QQ). For ease of viewing the results on standard screens, the width of QQ retrievals is limited to 79 characters per line for most tables. For tables with many columns, we have therefore selected those columns which we think will be most useful. QQ has two implementations: SQQ (Single Quick Query) is designed to query a single database and MQQ (Multiple QQ) provides the capability of searching many or all tables with a single specification of celestial position.

When the user needs columns not included in the two "quick query" options (above), "DBQ" allows the user to specify any filter condition on any column for search, and to specify which columns to retrieve. Although more complicated to use, this provides maximum flexibility for searching our data. Once built, the user may save these queries for future use.

4–4   Current status

The number of 'logins' to the EOLS continues to increase. Table 4–2 gives the average logins per month from outside the Center for Astrophysics.

Year	Average number/month
1989	44
1990	32
1991	76
1992	115
1993	219

Table 4–2: The Average number of logins per month

Since the first CD-ROM was mounted (2.5 years ago), more than 6000 HRI and IPC images and more than 1000 documents have been downloaded. We do not have an accounting system for files obtained via the Vax/VMS 'copy' command or for emailing of documents.

If we succeed in obtaining funding, there are several improvements we would like to implement. We need to obtain more reliable ROSAT pointing information; we need to provide precession to any epoch on retrieval; and we need to provide a method for users to make searches from a list of positions (*i.e.* 'batch processing'). This capability would also allow cross–correlation of entries between different tables so that one could use the result of a retrieval as the input for a subsequent search.

Acknowledgements

It is a pleasure to thank T. Karakashian who is responsible for most of the software and system architecture and H. Andernach who has supplied us with his collection of radio catalogues and helpful advice on a number of issues. From its inception until October 1, 1993, the EOLS was supported by NASA contract NAS8-30751.

Contact information

For additional information:

Internet	`einline@cfa.harvard.edu`
DECnet	`CFA::EINLINE (6699::EINLINE)`
UUCP	`...!harvard!cfa!einline`
BITNET	`EINLINE@CFA`

or contact

D. E. Harris   MS-3
Center for Astrophysics
60 Garden St.
Cambridge MA 02138, USA
Tel: +1 (617) 495-7148
Fax: +1 (617) 495-7356

Access Pointers

[→1] EOLS: `http://hea-www.harvard.edu/einline/einline.html`

5

The EUVE Archive

Center for Extreme Ultraviolet Astrophysics

The Center for EUV Astrophysics [→1] is part of the Space Sciences Laboratory at U.C. Berkeley.

5–1 Extreme UltraViolet Explorer Mission

NASA's Extreme UltraViolet Explorer (EUVE) satellite was launched on June 7, 1992 from Cape Canaveral, Florida on a Delta II rocket. The payload contains three EUV scanning telescopes equipped with imaging detectors as well as a Deep Survey Spectrometer instrument which divides the light from a fourth telescope between an imaging detector and three EUV spectrometers.

The first six months of the mission were dedicated to mapping the EUV sky with the scanning telescopes. The mission is now in the Guest Observer phase and will be extended through at least 1997.

5–2 EUVE Guest Observer Center

The EUVE Guest Observer Center [→2] provides information, software, and data to EUVE Guest Observers. It is located in the Center for EUV Astrophysics at U.C. Berkeley.

Information about Cycles I, II and III proposals, including PI names, Proposals and Targets are available on the EUVE WWW server. Processing for the third round of EUVE approved pointed observations is underway, and information about Cycle IV observations will be available soon.

Astronomers who are preparing proposals may find several utilities useful in preparing their proposals. It is not possible to specify an EUVE observation of a particular target on a specific day, however, if researchers have a tightly constrained observation, then the target visibility program can be used to verify that the target is observable. Given position and exposure this program will generate a suitability plot. A suitability plot describes how well

D. Egret and M. A. Albrecht (eds.), Information & On-Line Data in Astronomy, 43–46.
© 1995 *Kluwer Academic Publishers.*

EUVE can observe the target on any given day. The orbital elements program provides detailed information about scheduling an EUVE observation.

The ISM program computes the transparency of the Interstellar Medium using a commonly used model that only requires a few parameters. To determine the appropriate ISM parameters, an ISM Hydrogen Column Density Search Tool provides a set of reference ISM densities for a given position.

The EUVE Public Archive [→3] offers up-to-date data products from publically available EUVE data sets. Scientists are aadvised to carefully review the data sets available on-line before submitting proposals.

The Center maintains abstracts of papers using EUVE data. Astronomers who have published data relating to EUVE data should send a LATEX markup file or any other electronic format that astrophysical journals accept. The EGO Center will create an image of the abstract that is consistant with the document mark up.

5–3   The EUVE Science Archive

The EUVE Science Archive was established in order to archive and distribute EUVE-related material in an innovative and efficient manner. The EUVE Public Science Archive [→3] contains image and spectral data from EUVE as well as EUVE related preprints on line. Source fluxes and raw data may be requested via the World-Wide Web.

The Archive infrastructure includes:

▷ Mosaic

▷ gopher

▷ Anonymous FTP

▷ WAIS searches

▷ NASA Astrophysics Data System (ADS)

▷ CD-ROM Series

▷ Electronic Mail Server

▷ EUVE Electronic Newsletter

Laid upon this foundation are the contents of the site, which include a variety of EUVE-related data products (e.g. images, catalogs, maps, and spectra), software (e.g. interstellar medium flux attenuation code), documenfätion (e.g. newsletters and scientific and technical abstracts, papers and articles), and user services (e.g. electronic tools for previewing, processing and ordering data). Expansion of the Archive is an on-going process; new data are added as the relevant proprietary data rights expire, new tools and user services are continuously under development and additional modes of access are under investigation.

5–4   The EUVE Sky Surveys

The WWW pages provide access to EUVE sky survey data. EUVE was launched on 7 June 1992, from Cape Canaveral, Florida, into a near-Earth (550 km) orbit. Following seven weeks of in-orbit checkout, the survey phase of the mission began on 24 July and included the following:

  ▷ an all-sky survey conducted by the three "scanning" telescopes in four band passes covering the entire extreme ultraviolet range (60-740 Å);

  ▷ a concurrent "deep survey" (more sensitive by a factor of ten) of a 2x180 degree swath along the ecliptic conducted by the deep survey telescope in two band passes covering the 70-365 Å range.

The EUVE survey data products are of three distinct types:

CATALOGS —These are the published catalogs of detected and confirmed EUVE sources.

SKYMAPS —These are binned images of sections of the sky. The skymaps are useful for verifying "source" detections, measuring rough count-rates and viewing large areas of sky. They are quickly processed (hours), are easy to use and require no supplemental calibration data. The drawback to skymaps is that they are binned and contain no timing information.

PIGEONHOLES —These are files which contain the photon event information within a given radius of a given location on the sky. The pigeonholes are useful for calculating accurate source count-rates and, since they contain timing information, for constructing light-curves. The drawbacks of working with pigeonholes are that they are slow to process (days), are complicated to use and they require the use of supplemental calibration data.

The U.C. Berkeley science team at the Center for EUV Astrophysics (CEA) has made significant efforts to produce these catalogs, skymaps and pigeonholes, and to make available some ancillary software (*e.g.*, the services available below). Although the survey data is publicly accessible, due to computer resources and the time required to process incoming requests, CEA recommends that data be accessed in the following manner (quoted from the on-line server):

  1. Search the available EUVE source catalogs for your target(s) of interest. Since data for the brightest EUV sources has already been published, this step may be the quickest way to find what you need. (Note: A tool

to assist you in this search will soon be available.) If your target(s) of interest do not appear in the catalogs, continue to step 2.

2. Determine if your target(s) was (were) detected by EUVE by running the EUVE Count-Rate service. This service calculates and returns count-rate information by analyzing the skymaps to search for any significant source(s) near the input location. Count-rate results are processed within hours and sent via e-mail to the requestor.

3. After running the EUVE Count-Rate service in step 2 above, you should verify any interesting results by examining the portion of the skymap that contains the target(s) of interest. This is done by running the Skymap Request service (accessible at the bottom of the EUVE Count-Rate service form).

4. Once a source detection is confirmed via the skymaps from step 3 above, you may wish to retrieve the actual photon event list. To do this you must run the Pigeonhole Request service (accessible from the bottom of the Skymap Request service form).

NOTE FROM THE EDITORS

This chapter has been compiled by the Editors from the on-line WWW service [→1].

Contact Information

For more information on the Public Archive send mail to:

Dr. Carol Christian: `carolc@cea.berkeley.edu`

Access Pointers

[→1] Center for EUV Astrophysics: `http://www.cea.berkeley.edu`
[→2] EUVE Guest Observer Center:
`http://www.cea.berkeley.edu/EGO/EGOProgram.html`
[→3] EUVE Public Archive:
`http://www.cea.berkeley.edu/Archive/ArchiveHomePage.html`

6

The IUE Archives

M. Barylak, A. Talavera, W. Wamsteker, J. D. Ponz &
C. Driessen

6–1   The IUE Context

In the 1970s the US National Aeronautics and Space Administration (NASA), the UK Science and Engineering Research Council (SERC) and the European Space Agency (ESA) joint together to build, launch and operate the International Ultraviolet Explorer (IUE) satellite. This satellite was designed to enable scientists around the globe to study ultraviolet (UV) radiation.

Many of the astrophysically important atomic processes generate UV radiation (*e.g.* atomic Hydrogen at 1216Å). Invisible to the human eye, UV radiations spans from 1000Å to 3000Åand is absorbed substantially by ozone molecules in the Earth's stratosphere. Hence the need to use satellites outside the Earth atmosphere to study ultraviolet radiation coming from the stars and other celestial bodies.

The IUE satellite was conceived as a spectroscopic space observatory to collect primarily ultraviolet spectra of low (6Å) and high resolution (ca. 0.02Å) under an international collaborative project. The science instruments of IUE are a 45 cm Cassegrain telescope and the two spectrographs covering the range from 1150Å to 3200Å. SERC provided the on-board cameras with associated software. NASA provided the spacecraft with the rest of the scientific instrumentation and ESA was responsible for the solar panels.

The concept of IUE, to be used in much the same manner as a ground-based observatory dictated a high geosynchronous orbit about 36,000 km above the surface of the Earth. Two groundstations control the satellite 24 hours per day, *i.e.* the Goddard Space Flight Center (GSFC) in Maryland (USA) for 16 hours and ESA's satellite tracking station in Villafranca del Castillo (VILSPA) near Madrid (Spain) for the remaining 8 hours.

D. Egret and M. A. Albrecht (eds.), Information & On-Line Data in Astronomy, 47–55.
© 1995 *Kluwer Academic Publishers.*

6-2   Contents of the Archive

After 16 years in orbit, IUE has taken nearly 100,000 spectra of about 8000 unique astronomical sources. Three archive centers (one for each involved agency) maintain and distribute IUE data, i.e. National Space Science Data Center (NSSDC) at GSFC, the Rutherford Appleton Laboratory (RAL) World Space Data Center C1 in the UK and at VILSPA.

The *IUE Merged Observing Log* (*i.e.* the observing logs of both GSFC and VILSPA) holds important observing parameters of all archived images. It is available on-line at VILSPA since 1984 (Barylak, 1991). Currently, access is or will be provided via NASA's Astrophysics Data System (ADS, see Weiss & Good, 1991), the Wide Area Information Server (WAIS, Fullton, 1993), SIMBAD (Egret *et al.* , 1991, under measurements) and the World-Wide Web (WWW; White, 1993). The IUE Merged Log enables scientists to identify observations of their sources of interest. To facilitate this identification, the object nomenclature and coordinates are revised and homogenized annually in collaboration with the Centre de Données astronomiques de Strasbourg (CDS) (see Egret *et al.* , 1992). These data were also incorporated into the Uniform Low Dispersion Archive.

2.1   THE UNIFORM LOW DISPERSION ARCHIVE

For the first time in the history of Astronomy, a directly usable (because in physical meaningful units) spectroscopic data archive has been created and placed on-line, accessible through the general intercomputer networks available to most Astronomers. This is the on-line IUE *Uniform Low Dispersion Archive* (ULDA) (Wamsteker *et al.* , 1989).

ULDA is a compact subset of the IUE archive consisting of all low resolution (6Å) spectra taken with IUE in either of its two nominal wavelength ranges (1150–2000Å and 1825–3300Å) with any of the three cameras which have been and are still operational *i.e.* , the short-wavelength prime (SWP) and the long-wavelength prime and redundant (LWP/LWR) cameras. The data consist of spectra which are absolutely calibrated in terms of flux ($erg \cdot secs^{-1} \cdot cm^{-2} \cdot Å^{-1}$) and wavelength (Å) together with the standard IUESIPS data quality flags .

To optimize access and to minimize data handling, ULDA is distributed with its support software package (USSP) worldwide to about 25 national hosts serving some 900 scientists in 35 countries. At a time, where networking was starting, ULDA/USSP offered a pioneering concept in data distribution: any scientist can help himself interactively to search for, to select and to downlink IUE low dispersion spectra to his own computer.

The use of the system has been kept simple and the software is quite easy to adapt to different systems through the use of strictly controlled standards.

The overall concept of USSP is specified by the following requirements:

> ▷ the spectra must be accessible to scientists in many institutes spread over a number of countries;

> ▷ all users must have the *same* data available to them, *i.e.* any updates and additions to the ULDA should be accessible to all users;

> ▷ retrieval of spectra must be both fast and easy to perform, preferably coming as close as possible to having the spectra on line;

> ▷ it should be easy to use;

> ▷ the data should be available for manipulation independent of the specific data analysis software available to the individual scientist;

> ▷ no intermediaries between the scientist and the ULDA (*i.e.* self-registration without intervention by IUE Project);

> ▷ crash tolerant;

> ▷ the system looks after itself wherever possible. (*e.g.* workfiles);

> ▷ the program language must be commonly available, *i.e.* the system is not portable but *easily* adaptable.

This low resolution data set satisfies nearly 30% of all data retrieval from the IUE archive which distributes around 56,500 spectra annually (see table 6–1). Considering that the demand for IUE observing time exceeds the available time by a factor of 3, the IUE archives and its data distribution constitute an important part of the IUE Project.

The current version of ULDA/USSP (*i.e.* Version 4.0) gives access to ca. 54,000 absolutely flux calibrated low dispersion spectra.

The ULDA/USSP is a *distributed system* with VILSPA as its Principal Centre. Therefore the system can grow without affecting the overall performance. The strategic copies of ULDA at the national host have increased the availability of IUE low dispersion data. The following table gives an estimate of the different IUE data dissemination activities:

2.2 THE IUE FINAL ARCHIVE

The IUE project is generating a new version of the archive —the so-called *Final Archive*— while experience is still at hand. The main reasons are: Firstly, the reduction method has evolved during the lifetime of the project and different sets of calibrations were used, therefore, it is difficult to combine low and high resolution data or to compare observations taken several years apart. Second, experience and knowledge about the instrument have been developed in the users community. The appropriate feedback from the

	Annual rate	Total nr.	% of total
6 yrs ULDA/USSP	24,300	146,000	29.6
10 yrs NSSDC/RDAFs (USA)	27,000	270,000	54.6
15 yrs VILSPA	3,500	53,000	10.7
15 yrs RAL	1,700	25,300 ·	5.1
Total:	56,500	494,300	100.0

Table 6–1: Number of delivered IUE spectra - Nov. 1994 (not including original Guest Observer data delivery).

IUE community allowed the project to identify new numerical methods that substantially improve the quality of the reduced data.

The basic differences between the old and the new processing schemes are:

▷ New set of Intensity Transfer Functions (ITF) to linearize the response of the cameras. The ITFs are defined in the 'raw sampling domain' to avoid smoothing.

▷ The geometric transformation required to align raw images with the ITF is defined by a set of vectors, derived from the fixed pattern noise in the images rather than using the 'fiducial marks'.

▷ The resampling of the linearized image into a uniform rectified domain is done in a single step. The transformation includes

  1. displacements between the raw image and the ITF domain,
  2. rotation to align horizontally the spectrum,
  3. aperture alignment and
  4. wavelength linearization.

▷ The new extraction algorithm uses a weighted slit. The extraction weights are based on a noise model, derived empirically.

▷ The inter-order background of high dispersion spectra is estimated by a polynomial fit, based on the regions of the images with no spectrum.

The improvement in the S/N ratio is about 10-30 % for individual spectra that can be as high as 50 % in the case of under-exposed spectra.

The absolute flux scale is defined by the IUE standard stars. New observations of White Dwarfs are used to obtain the relative sensitivity of the cameras, by comparing them with model atmospheres. The time-dependent sensitivity degradation is included in the final calibration.

The Final Archive contains the raw data, extracted spectra, intermediate data sets and the observing and processing parameters stored in FITS format.

The estimated size of the archive is 8 Mbytes per low dispersion observation and 11.4 Mbytes for high dispersion (see Tables 6–2 & 6–3).

Description	Size	Mbytes	FITS type
Raw Image (RI)	768×768	0.6	image
Linearized Image (LI)	768×768	2.4	image
Vector Displacement (VD)	2×768×768	4.8	image
Resampled Image (SILO)	640×80	0.24	image
Extracted Spectra (MXLO)	n/a	0.02	table
Total (with header for each file)		ca. 8.2 Mb	

Table 6–2: Planned contents of the IUE Final Archive – *Low Dispersion*

Description	Size	Mbytes	FITS type
Raw Image	768×768	0.6	image
Linearized Image (LI)	768×768	2.4	image
Vector Displacement	2×768×768	4.8	image
Resampled Image (SIHI)	768×768	2.4	image
Extracted Spectra (MXHI)	n/a	1.2	table
Total (with header for each file)		ca. 11.4 Mb	

Table 6–3: Planned contents of the IUE Final Archive – *High Dispersion*

6–3   Usage of and Science with IUE

The IUE archive because of its large volume and scientific value will certainly be used for many years to come. This is also indicated by the current usage and data retrieval statistics (Table 6–1).

Of the 131,490 hours available to IUE over the years 1978 to 1992, 45,565 hours were used in exposing the SWP camera. The LWP camera is next with a total exposure time of 17,124 hours. The LWR which developed a flare around 1983, is no longer being used on a routine basis and was exposed for 10,676 hours.

The scientific accomplishments of IUE are reviewed *e.g.* by Kondo *et al.* (1987), Kondo *et al.* (1988) and Wamsteker (1994).

3.1   IUE BIBLIOGRAPHIES

More than 2,400 refereed papers can be found in major astronomical and geophysical journals enunciating a large variety of astrophysical problems. The number of refereed papers have been increasing with time and demonstrate amply the scientific productivity of IUE.

A complete index of IUE related publications has been published recently in the IUE NASA Newsletter (Pitts, 1993).

3.2   ATLASES AND CATALOGUES

A list of atlases and catalogues based on IUE data has also been compiled (see Egret *et al.* , 1991 and Egret *et al.* , 1992). All atlases are listed again in the references for your convenience (Refs. [7] till [18]). More than 40 major atlases and catalogues have been published based on IUE data. Of general interest are the series of IUE ULDA Access Guides (see references) which were published on such diverse topics as Dwarf Novae and Nova-like Stars, Comets, Normal Galaxies and Active Galactic Nuclei. Recently, Wu et al. (1992) have compiled a very comprehensive atlas of selected astronomical objects. A complete list of the available atlases and catalogues is presented by Pitts (1993).

3.3   LONG-TERM MONITORING

About 2,000 objects have at least 2 or more IUE observations in different years over the life time of IUE (see [13]). Table 6–4 lists a sample of objects for which detailed UV variability studies can be carried out. Objects observed during all the 15 years include the major planets, IUE standard stars, Active Galactic Nuclei, PMS stars, Cataclysmic Variables, Novae, etc.

6–4   Future Plans

As mentioned above the IUE Final Archive will contain documented (via FITS headers) raw, calibrated and reduced data. The FITS headers will hold all information necessary for both scientific interpretation (including homogeneous object identifications and coordinates) and data reduction. The final data storage of the IUE Final Archive are optical disks (6.5 Gbytes) in FITS format.

The IUE data distribution system is currently being defined in a phased approach that incorporates the experience with ULDA/USSP, using modern technologies in both networking and user interfaces. The definition of this system will be done in collaboration with ULDA host managers in such a way that experience and ideas from the users community are incorporated at an early stage of the design.

9		10	11	12
Callisto	RW Aur	3C 120	AM Her	3C 273
Europa	R Aqr	AG Dra	Cyg Loop	3C 390.3
GK Per	RX Pup	HD 149757	HBV 475	HD 210839
HD 120324	SS Cyg	HD 150798	HD 108903	HD 24912
Arcturus	CH Cyg	HD 186791	HD 162732	HD 5394
HD 128620	EX Hya	HD 209750	HD 164284	Neptune
HD 193793	Z Cam	HD 222107	HD 22049	RR Tel
HD 200120	3C 446	V711 Tau	HD 41335	
HD 203467	3C 382	HD 33328	FY CMa	
HD 32068	AKN 120	HD 34816	Procyon	
MKN 509	BP Tau	HD 37490		
NGC 3783	BD +48 1777	HM Sge		
NGC 4151				

Table 6–4: Sample of objects observed at least in 9 to 12 different years (taken from Pérez, *et al.* , 1994)

In summary, the main design aspect is to structure the system in three levels, corresponding to the Master Node, several National Host nodes and an unlimited number of end-user nodes; very similar to the ULDA/USSP distribution architecture. End users will access National hosts via high level Internet services to query the catalog and request data available either at the National host or provided by the Master node. Details of the design were discussed during the ULDA Host Manager meeting at VILSPA in June 1994. A prototype of the system under Mosaic is available at VILSPA to access Final Archive data required for production Quality Control.

Plans are in hands which will regroup parts of the IUE archive on a medium such as the CD-ROM. This new medium will avoid on-line access to data centres especially in cases where large amount of data are retrieved and in countries without reliable telecommunication facilities.

References

[1] Aiello, S., et al.: 1988, 'Atlas of the Wavelength Dependence of UV Extinction in the Galaxy', *Astron. Astrophys. Suppl.*, **73**, 195.
[2] Barylak, M. 1991: 'The VILSPA Database Users Guide', *IUE ESA Newsletter*, No. **37**.
[3] Benvenuti P. et al., 1982: 'An Atlas of UV Spectra of Supernovae', *ESA SP-1046*.
[4] Doazan, V. et al., 1991: 'A Be Star Atlas of far UV and Optical High-Resolution Spectra', *ESA SP-1147*.

[5] Egret, D., Wenger, M., Dubois, P., 1991: 'The SIMBAD astronomical database', in *Databases and On-line Data in Astronomy*, Kluwer, 79.

[6] Egret, D., Jasniewicz, G., Barylak, M., Wamsteker, W., 1992: 'Homogenization of the nomenclature of the IUE log of observation', Proceedings of *Astronomy from Large Databases II*, Haguenau, ESO proceeding **43**, 265.

[7] Feibelman, W.A., et al. 1988: 'IUE Spectral Atlas of Planetary Nebulae, Central Stars, and Related Objects', *NASA Reference Publication #1203*.

[8] Fullton, J., 1993: 'WAIS' in *Intelligent Information Retrieval: The Case of Astronomy and Related Space Sciences*, Kluwer, 113.

[9] Heck, A. et al., 1984: 'IUE Low-Dispersion Spectra Reference Atlas, Part I: Normal Stars', *ESA SP-1052*.

[10] Kondo, Y., et al., 1987: *Exploring the Universe with the IUE satellite*, D.Reidel, Vol. **129**.

[11] Kondo, Y., Boggess, A., Maran, S. P. 1988: *Annual Review of Astron.& Astrophysics*, **27**, 397.

[12] LaDous, C., 1989: A Catalogue of Low-Resolution IUE Spectra of Dwarf Novae and Nova-like Stars,*Space Science Reviews*, **49**, 425.

[13] Pérez, M.R., Thompson, R.W., Barylak, M., Bonell, T.J., 1994: 'The Evolution of the IUE Archive' in *Proceedings of Frontiers of Space and Ground-based Astronomy*, Kluwer, 1994, in press.

[14] Pitts, P.S., Imhoff, C.L., 1989: *IUE NASA Newsletter*, **37**, 1.

[15] Pitts, P.S., 1990: *IUE NASA Newsletter*, **41**, 49.

[16] Pitts, P.S. 1993: *IUE NASA Newsletter*, No. **50**, 35.

[17] Rocca-Volmerange, B., Guiderdoni, B., 1988: An Atlas of Synthetic Spectra of Galaxies, *Astron. Astrophys. Suppl.* **75**, 93.

[18] Rosa M., et al., 1984: 'IUE UV Spectra of Extragalactic H II Regions. I, The Catalogue and the Atlas', *Astron. Astrophys. Suppl.* **57**, 361.

[19] ULDA-Guides:
LaDous, C. 1989, *IUE-ULDA Access Guide*, No. **1**, ESA SP-1114.
Festou. M.C., 1990 *IUE-ULDA Access Guide*, No. **2**, ESA SP-1134.
Longo, G., Capaccioli, M. 1992, *IUE-ULDA Access Guide*, No. **3**, ESA SP-11524.
Courvoisier, T.J.L., Paltani, S. 1992, *IUE-ULDA Access Guide*, No. **4**, ESA SP-1153.

[20] Van Steenberg, M., Green, J. L. 1991, in *Databases & On-Line Data in Astronomy*, p. 151.

[21] Walborn, N.R., et al., 1985: 'IUE Atlas of O-Type Spectra from 1200 to 1900 Å,*NASA Reference Publication #1155*.

[22] Wamsteker, W. et al., 1989, *Astron. Astrophys. Suppl.* **79**, 1.

[23] Wamsteker, W., 1994: Proceedings of Frontiers of Space and Ground-based Astronomy, Kluwer, *in press*.

[24] Weiss, J.R., Good J.C., 1991: 'The NASA Astrophysics Data System' in *Databases & On-line Data in Astronomy*, Kluwer, 139.

[25] White, B., 1993: 'WorldWideWeb (WWW)' in *Intelligent Information Retrieval: The Case of Astronomy and Related Space Sciences*, Kluwer, 127.

[26] Willis, A.J., et al., 1986: 'An Atlas of High-Resolution IUE Spectra of 14 Wolf-Rayet Stars', *Astron. Astrophys. Suppl.* **63**, 417.

[27] Wing, R.F., 1983: 'Atlas of High-Resolution IUE Spectra of Late-Type Stars, 2500 till 3230 A', *Perkins Observatory Special Publication #1*.

[28] Wu, C.C., et al., 1991: 'The IUE Ultraviolet Spectral Atlas', *NASA IUE Newsletter*, **43**.

[29] Wu, C.C. et al. 1992: *NASA Reference Publication #1285*.

Access Pointers

[→1] Vilspa WWW server: http://www.vilspa.esa.es/

[→2] IUEDAC (NASA):
http://iuesu1.gsfc.nasa.gov/iue/iuedac-homepage.html

7

Building the archive facility of the ESO Very Large Telescope

Miguel A. Albrecht, Michèle Péron & Benoît Pirenne

7-1 Introduction

The experience gained while building a science archive for the NTT (Albrecht & Grosbøl, 1992) has shown that the main problems lie less in solving the technical issues than in making the observatory and the community aware of the advantages of a *traceable* and *predictable* telescope operation. Traceable operations allow to quickly understand the source of problems and is the basis for high-quality, detailed scientific analysis of the data. Predictable operations allows for good observing programme preparation and is the basis for high efficiency.

Traceable and predictable telescope operations imply that:

  ▷ a comprehensive data definition (file structure and headers) is available, and changes to it are subject to a configuration control procedure;

  ▷ standard procedures for instrument configuration and calibration are pursued and documented;

  ▷ quality and performance parameters are constantly monitored;

  ▷ instrument and operations manuals are maintained up-to-date;

  ▷ all relevant operations and atmospheric conditions are recorded;

  ▷ an archive keeps track of events and provides a long-term storage of both data and information.

Although most of these points were present in the classical operation of the NTT, it was not until recently (Baade *et al.*, 1994), that ESO undertook a systematic approach to tackle them one by one. The *NTT Upgrade Project* has since then served as a constant source of learning and has been a valuable experience —the most drastic phase yet to come in 1996, when the complete telescope control soft- and hardware is replaced with the new VLT-standard products.

D. Egret and M. A. Albrecht (eds.), Information & On-Line Data in Astronomy, 57–71.

Needless to say that a traceable and predictable telescope operation is essential for a useful science archive. But it is also important to note that once such an operational setup has been achieved, the effort necessary to build the archive is small and can be easily accommodated within the overall data management activities.

The NTT archive proved to be a valuable resource in the assessment of data quality and operation stability (Peron *et al.* 1994) and in the formulation of solutions (Albrecht & Benvenuti, 1994). From this experience, we set out to define and design the archive facility for the ESO Very Large Telescope. For the part of this facility that will operate "on-line" at the telescope, this definition is embedded in an ESO wide effort aimed at the analysis of data and information flow through the complete data system. Undertaken by the so-called *Data Flow Working Group*, this effort should deliver by mid-1995 an overall data system design. Object models, dynamical models and functional diagrams are among the most important results expected from this activity, in which most ESO divisions and groups participate actively.

Although preliminarily, we will describe in this chapter the issues that need to be addressed when building the VLT Archive: they include the data interface definition, on-line tools and facilities, the science archive, the engineering archive and a general purpose catalog and database browser.

7–2   Data interface definition

One of the major lessons learned from the archive experience with the NTT, is the strong need to define and manage a dictionary of data structures and formats that permeates all aspects of the data system from observation preparation, data acquisition, data reduction, quality control and the archive.

As opposed to a space experiment where the instrumental configuration remains stable for the lifetime of the mission (part of the mission in the case of HST), instruments on the ground are subject to permanent change (the mounting of new filters can be a daily procedure). Also, in the course of trouble–shooting procedures it is often required to augment the amount of information being recorded by a few more parameters and/or status values. Therefore, it is essential that changes and additions to the data dictionary are coordinated throughout the data system such that when *e.g.* a header keyword is added or a new filter name is defined, it can be immediately used by the reduction software, become a new index in the archive database and be properly interpreted by an instrument trend–analysis tool.

Another lesson learned was the need to record information differently according to different contexts and serving different purposes. For instance, the astronomer needs to have the width of a slit included in the frame header and coded in seconds of arc, while an instrument engineer will need to know the encoder value, the width in micron and the conversion formula used to translate the values from micron to seconds of arc. The approach taken for the NTT archive: "record only engineering (not derived) parameters in SI units" proved to be insufficient, the headers were not easily readable and became unnecessarily long, including sometimes duplicated information.

Also, in many cases it was considered desirable —in particular in user interfaces— to be able to *alias* a meaningful name to a possibly cryptic keyword, *e.g.* FILTER instead of EMRFILT.

In an effort to address these issues, ESO has setup a Data Interface Control Board (DICB) . The main purpose of the board is to promote the standardization of, and enforce configuration control on data structures that are

1. used by ESO to deliver data products of any kind to the community
2. needed by users of ESO facilities to submit observation related information to the organization.

These data structures are defined and managed through a *Data Interface Control Document* and are subject to a configuration control procedure.

The *data interface* itself covers data and file formats, naming conventions, meanings and physical units when applicable. Examples include:

▷ the definition of all FITS headers of exposures acquired with ESO telescopes. In particular the keyword names (and possibly aliases), their meaning, format and physical units of their values when applicable;

▷ the definition of data structures that represent and report events in the operations log. In particular the format of the log, name of events, their meaning, format and physical units of their values when applicable;

▷ the definition of data structures needed for calibration purposes. In particular the format of tables, lists and other data collections describing instrumental characteristics, laboratory and astrophysical data, names of elements, their meaning, format and physical units of their values when applicable;

▷ the definition of data structures applied to store data that has been calibrated using ESO standard reduction procedures;

▷ the definition of formats for target and instrument setup information as part of an observation preparation phase.

This effort has brought together representatives from all relevant groups in the organization and has been very successful in the definition of data products and the coordination of related activities.

As a first result, the data products (*i.e.* header and log keywords) of the acquisition system have been completely reviewed. The overall structure has been redefined such that

▷ science headers contain only information that is relevant to data reduction and analysis and is recorded in end-user units such as arc seconds for slit widths, *etc.*;

▷ science headers keep the non-standard hierarchical keywords (see Albrecht & Grosbøl, 1992), but a tool is provided to translate headers from one semantical specification to another (*e.g.* hierarchical keywords with ESO units, names, *etc.* to *e.g.* IRAF–STSDAS naming/unit conventions); the translation is table driven, the table itself can be provided by the user or by the observatory;

▷ a number of log files record all information relevant to science operations; in particular, an *operations log*, a *configurations log*, a *standard–reduction log* and a *conditions log* (see also section below);

▷ an *observation summary table (OST)* provides a user–friendly view of the data harvest both at the telescope and at home during post-observation data analysis (see next section).

The description of the keywords is defined via a set of data interface dictionaries, each one concerned with a given context such as a particular instrument. The board has released a specification for the dictionary itself and is currently releasing the associated *Data Interface Control Document*.

This document includes the specification of the VLT *generic* data interface dictionary. The generic dictionary lists mandatory and optional keywords for all contexts (telescope, instrument, detector, *etc.*) and specifies the rules for defining keywords within specific dictionaries, *e.g.* of a particular instrument.

The document also includes the definition of an alias scheme as a vehicle to provide alternative, user-friendly names for elements that are planned to be used throughout the data system. An example of such an alias is `RedArm.Filter` instead of `HIERARCH INS OPTI3 NAME` or `EMRFILT`. Aliases will provide a standard access to instrumental elements throughout the data life-cycle, for instance while preparing observations with instrument information sytems, specifying the instrument setup, reducing the data or while querying the archive database. The alias translation table for each context will be an appendix of the corresponding data interface dictionary.

It is expected that the ESO DICB will gradually cover in the coming year all aspects of the data interface as listed above.

7–3   On–line archive tools and databases

The current architecture of the control software foresees the following functions to be performed by the on–line archive system:

- ▷ organization and storage of the data harvest;
- ▷ organization and storage of the calibration database;
- ▷ management of data delivery to observers;
- ▷ transfer of data and logs to the off–line archive.

The distribution of functions in this area is subject to the results of the *Data Flow Definition working group* which is expected to finalize the overall system architecture by mid 1995. The description below represents therefore only a snapshot of the system design at the time of writing (December 1994).

In order to implement the functionality within an operational setup (maximum fault tolerance), a design was adopted that would allow for tasks to run independently and in parallel. The figure 7–1 shows the draft architecture of the on–line system.

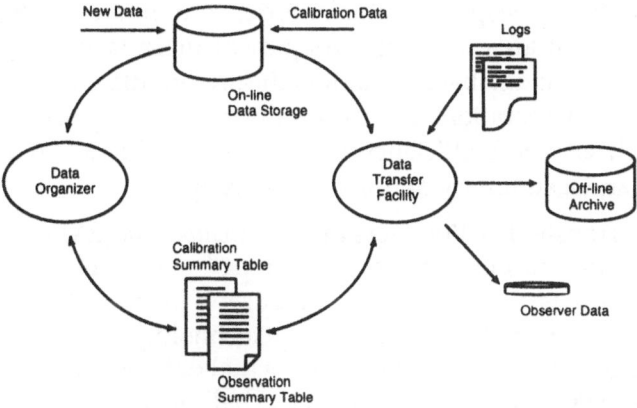

Figure 7–1: Draft architecture of the on-line Data Management Facility.

3.1 THE DATA ORGANIZER

Single exposures, as they are delivered by the data acquisition system, are read and checked for consistency by the *Data Organizer* (DO) (Peron *et al.* 1994), a tool developed to maintain an accurate book–keeping of all exposures of a given programme. The DO keeps an Observation Summary Table that contains for every exposure all relevant keywords from the header together with information required to group science and calibration frames, user notes, quality flags, *etc.*

When new calibration frames are obtained, either as raw (*e.g.* new BIAS) or processed calibrations (*e.g.* averaged BIAS delivered by the standard reduction pipeline), the DO adds a corresponding record to a Calibration Summary Table (CST). The CST represents the latest state of the calibration database

and includes the best available calibration data at any given time. Both tables CST and OST are used by the pipeline reduction procedures to calibrate
major instrument modes in a standard way.

In addition to this unattended operation, the DO features a user interface
to allow observers and science operators to view and manipulate the OST and
CST. Among other functions, the DO offers a tabular view of selectable header
keywords, a utility for classifying files into user–definable categories and for
defining associations between categories, a possibility to make annotations
to frames, *etc.*

3.2   DATA TRANSFER

The *Data Transfer Facility* includes background tasks to collect and transfer
data, tables and logs to the off-line archive storage. Most of these tasks are
activated by the telescope operator in order to minimize interference with
the telescope local area network. As part of these tasks, the observation
catalog and archive database are automatically populated. Also, on a night
by night basis, an operations *digest* is produced including science operations
statistics such as overall efficiency, telescope vs. ambient conditions (meteo
and seeing) and a summary of unforeseen events.

The Data Transfer Facility offers the *File Archive Tool* (FAT) to observers for
managing the generation of the "observer tape" —the data delivery medium.
The name 'tape' is used here only because of historical reasons, in fact,
many developments in the area of storage technology (see chapter 23 in this
volume) suggest that most probably the actual data delivery medium in the
VLT era will be CD-ROMs rather than *e.g.* helical scan tapes. The FAT will
include functions to a) translate headers and tables from the ESO standard
FITS keyword definition to a user defined format (*e.g.* MIDAS, IRAF, *etc.*)
and b) store the user data on the delivery medium and/or copy data to other
destinations.

Both user interfaces for the DO and FAT will feature suitable options for
printing and saving the lists of items (*e.g.* in an ASCII file).

3.3   THE CALIBRATION DATABASE

In order to perform a reasonable quality standard reduction procedure on
data acquired in major instrument modes (ESO Science Operations Plan,
1994), a database of the most suitable calibrations available for a given mode
will be maintained on-line.

This database is populated with a) user calibrations as acquired within
the normal execution of the user's observing programme; b) observatory
calibrations obtained in regular intervals to complement the calibration pool

and to perform instrument trend analysis; and c) reduced calibration data produced by the standard reduction pipeline (*e.g.* averaged BIAS frames, *etc.*).

The main user of the calibration database will be the reduction pipeline. However, since calibration data is not subject to proprietary rights owned by observers (as is science data), the contents of this database is open to the community at large. Because the calibration database also contains the results of trend–analysis and performance measurements, it will always be possible to judge the quality of the data available for calibration purposes. Hence, some calibrations may be reusable, promoting this way a more efficient utilization of the observing facilities.

7–4   The Archive Facility

The stick by which the success of the VLT science archive will be measured is long. The IUE archive delivers to the community about 5 times more data per month than the satellite itself after 14 years of satellite operations. The HST archive is delivering, after 4 years of HST operations, about the same amount of data than HST itself. The goal is that the VLT science archive should, by the end of the third year of regular operation, deliver half as many observations as it receives.

In order to reach this goal, the archive facility must demonstrate that it can support users to:

▷ query and browse an *observation catalogue* by scientific criteria;

▷ understand the scientific goal which the original observation was obtained for;

▷ assess the quality of the data;

▷ retrieve all the information which describe the particular conditions (e.g. weather and seeing) under which the observation was obtained and the specific procedures (e.g. positioning of the telescope) which were used;

▷ retrieve all calibration files and information which are applicable to the specific observation, including trend and performance analysis of the instrument;

▷ retrieve information on the best suitable calibration algorithms and procedures;

▷ calibrate the data upon request using standard recipes;

The concept design of the archive facility can be best understood in terms of its components, the science and the engineering archives. This conceptual separation corresponds to the need to meet the requirements of two rather different user groups: the archival researchers and VLT science operations staff.

In the following sections we will address these requirements in detail.

4.1   THE SCIENCE ARCHIVE

The science archive includes:

THE OBSERVATIONS CATALOG defines the observations performed in terms of their scientific target, being these observed objects or phenomena. It represents the accumulated knowledge of "what" has been observed as opposed to "how" observations were done.

> The observations catalogue includes the PI–defined classification of targets, objects coordinates and aliases and references to other information related to the target like "ROSAT–source", *etc.*

> The observations catalogue will be cross–checked with existing data, *e.g.* the SIMBAD database (see page 166 in this book).

THE ARCHIVE DATABASE defines all information needed to characterize a specific exposure. It includes all instrumental characteristics and setup parameters used for a particular observation. Each frame in the database is associated to both "engineering" and "astronomical" quantities. "Engineering" parameters include encoder values, physical slit widths *etc.*, while "astronomical" ones are computed from the former to derive astronomically meaningful quantities such as slit widths in arcsecs, central wavelengths, *etc.*

> For each observation, a *preview* data set will be produced via a standard reduction recipe and stored on–line in the database. Usually, *preview* data is of limited quality and meant only as a "quick–look" presentation of the actual science data.

The science archive can be queried on–line via a catalog/database browser (see section below) which supports seamless navigation in between the observations catalog and the archive database.

4.2   THE ENGINEERING ARCHIVE

The engineering archive includes:

THE LOG DATABASE defines all information that characterize the environment in which a specific observation was done. It represents the logbook of telescope operation. It uniquely associates a scheduled observing programme to a set of acquired exposures.

> The log database includes night reports and the following log–files:

    ▷ the operations log: records all major operations performed and their results (*e.g.* telescope presets, instrument operations, detector readouts and possible preprocessing);

    ▷ the configuration log: records all changes in the overall configuration such as pointing models, mounted filters, adaptive and active optics parameters, *etc.*;

    ▷ the conditions log: records main meteorological an seeing measurements both ambiental and within the dome;

    ▷ the pipeline log: records all standard reduction steps applied and results obtained.

INSTRUMENTATION DATABASE contains all features required to characterize an instrumental setup. It includes all information about telescopes, instruments, detectors, *etc.* describing the history and present status of hardware in operation at the observatory. It includes static data such as filter transmission curves, grating and grisms efficiency curves, detector responses, but it is also fed with trend analysis results such as CCD linearity measurements and therefore reflects at any given time the performance status of all instrumental equipment. The database is updated via the configuration log and by trend analysis tools.

ENGINEERING CALIBRATION DATA includes all data acquired while performing the engineering calibration of optical and mechanical elements. Examples are data and parameters used to derive a new pointing model, raw data of CCD linearity measurements, *etc.*

The engineering archive can be queried on–line via a catalog/database browser (see section below) which supports seamless navigation in between the observation log and the instrumentation database.

4.3 STORAGE AND OPERATION REQUIREMENTS

The VLT will produce large amounts of data. The current estimate for all four telescopes is of the order of 5 Terabytes per year when operated separately and at least three times as much when operated in interferometry mode. With such data flow, the question of storage and retrieval becomes non-negligible. We can count on our side the fast pace of development currently observed in storage technology (see Pirenne & Durand in this volume, page 243), however, we must provide enough resources to keep an archive operation such that the facility remains usable (with *e.g.* short retrieval times) even after many years of operations (resources required grow with data volume).

Clearly, the solution to this problem must concentrate on the deployment of technical instead of human resources, *i.e.* ESO cannot afford to add one archive operator (or cost equivalent) per year of operation. But, will technology provide these solutions in time? Recent developments show very encouraging results: with the advent of multilayer optical storage, as announced for CD-ROM's for 1996, we will be able to store 6 GB of data on one single compact disk. With the already available and inexpensive use of Jukeboxes (auto changers) we can put on-line 500 disks (3 TB) per device at a cost of roughly US$ 25,000 (December 1994 figures). With this cost level, we can indeed easily afford putting the whole of the VLT archive on-line for many years to come!

7–5   The Catalog/Database Facility

Following the current Science Operations Plan for the VLT (ESO VLT Science operations Plan, 1994), there will be a strong need for flexible and user friendly set of tools for the preparation of observations. These tools will cover functions such as finding chart generation, instrument simulation and setup, guide star selection (and reference star selection for adaptive optics). On the other side, there is a need for a general purpose catalog and database browser that can provide user friendly access to both the science and engineering archives.

In this section, we describe the user requirements of the Catalog/Database Facility for the ESO VLT. This facility is meant to provide support to observation preparation tools and offer a database browser.

The approach taken here for the definition of requirements is to formulate them through realistic usage scenarios. To a large extent, the requirements described here reflect already existing functionality available in existing software (*e.g.* STARCAT, MIDAS, xephem). The aim of this description is to encourage the discussion of functions to become available in next generation of tools for observation preparation. At the same time, the lack of resources makes it imperative to approach such development in a collaborative manner and make this facility of use for a larger community.

The Catalog/Database Facility will be developed jointly by ESO, the Space Telescope European Coordinating Facility (ST–ECF) and the Canadian Astronomy Data Centre (CADC). The CADC is in charge of developing the Data Management Facility for the Gemini telescopes.

The design and development philosophy to be followed will be that of maximum reusability, which implies the integration of existing software as far as possible.

Typical applications supported by this facility are: a target acquisition tool and finding chart generator, a general catalog and database browser, a catalog cross correlation tool, a guide star selection server, *etc.*

From the system design point of view, such a facility is made of a number of different components: there is the actual data (*e.g.* standard catalogs, digitized sky survey), the storage engines (*e.g.* a DBMS) and applications and their corresponding user interfaces (see above). It is essential that the software is designed such as to allow (and encourage) the development of other applications than those originally envisaged. This implies that it must be "easy" to add new software object types and to develop the corresponding functionality. For instance, adding a new type of storage engine should be transparent to user applications. Ideally, we want to allow external new packages to be (re)used within our environment and also to allow external groups to develop applications (re)using our code.

An analysis of these requirements yielding design constraints is given in [2].

SCENARIO 1: DB BROWSING

User browses through catalogs and/or archive databases; narrows or broadens the selection by specifying constraints on some fields. Items of interest are chosen from a list of available columns. The display of the selected rows can be chosen as full, table or graph (*e.g.* a sky map or a plot) with a choice of suitable options. The view of items can be changed according to a *presentation* suitable for each data type, *e.g.* the number of decimal digits for floats, the length of a string or the format for RA and DEC (decimal degrees or HH MM SS.SSS). Data *transformation* options are also available for each data type, such as to support *e.g.* precession of coordinates, unit transformation. Displayed or selected data (*e.g.* mouse selected rows) can be saved on a file with suitable choice of formats (*e.g.* ASCII or FITS for tables, PostScript for graphs) or printed. Error checking is provided when entering qualifications on fields. On–line help and context sensitive documentation is available. Preview data of catalog records is visualized upon request, if available. Preview data could be one or more images, spectra or other data sets (*e.g.* a light curve) — the display tools will offer suitable choices to handle several previews. Requests for off–line retrieval of archive data is supported via a suitable tool.

User queries databases physically located on different sites and possibly running on different hardware platforms. The common denominator is only that all, user and servers are connected to the Internet. This consideration applies to all scenarios. Catalogs can be stored on different query engines (*e.g.* an RDBMS like Sybase, as an ASCII, MIDAS or FITS table).

SCENARIO 2: GUIDE/REFERENCE STAR SELECTION

A) Telescope control software (TCS) accesses a server in real time to obtain suitable guide stars for a given field of view. Constraints are specified via qualifiers on a standard set of catalogs (GSC-II, Hipparcos, Tycho, PPM). A visibility mask can be attached to the field of view (a ring describing vignetting and detector visibility constraints) *e.g.* the user needs a guide star between 2′ and 3′ from the galaxy for active guiding. The system would return a list of stars within the ring. TCS expects an answer in the sub-seconds range.

B) User invokes the 'StarSelect' widget from an application. Widget allows selection of one or more catalogs (from remote or local server). Application gives widget the field of view and visibility mask (see above) according to telescopes and instrument configuration. Widget displays star field plus mask and accepts star selections via a mouse button click. Another mouse button displays information about the star. Widget supports zoom, pan, flip and scale functions. Print and Save are available with suitable options. A choice or combination of target acquisition functions will be offered such as: position slit(s) on field, define detector window on field according to given criteria (*e.g.* include/exclude reference star, *etc.*). A list of selected guide stars can be saved on disk (local server).

Scenario 2–b can take place either at the telescope or at the user's institution in the context of observation preparation.

In both scenarios the actual selection of guide or reference stars can be driven by a list of pointings provided by user. A choice or combination of standard selection criteria will also be offered such as: find brightest guide star or most isolated guide star, find star with color similar to target or guide star closest to target (within a field).

The local server supports the management of user tables. These may be stored in FITS or ASCII tables.

SCENARIO 3: CATALOG CROSS CORRELATION

User cross correlates a private catalog (*e.g.* a target list) with a catalog on a server. Typical correlation criteria will be vicinity in RA/DEC space with constraints in mag and other fields, however, the cross-correlation function might also be used to *e.g.* find closest reference star to given target.

The user list may have been obtained by extracting a subset catalog from a server, by merging and/or editing lists from different sources, by simply typing in targets with an editor or a combination of the above. The local server supports the management of user tables. The input target list is available on a server. The result of the correlation is stored in a table in the local server.

User can start cross correlation tasks in parallel and therefore effectively correlate a target list against many standard catalogs. The cross-correlation of standard catalogs against another can be accomplished by first extracting regions of interest from one catalog onto local tables (see scenario 2) and then cross-matching these with another standard catalog.

SCENARIO 4: NAVIGATION AND ACCESS TO EXTERNAL RESOURCES

User wants to access the instrument information system (based on http) while browsing through an archive database, *e.g.* see filter description or instrument specifications of those involved in a particular observation within the log.

User browsing the instrument information system via a WWW viewer wants to access the calibration database on a server, *e.g.* see the last linearity measurements of a particular CCD.

SCENARIO 5: FIND SIMILAR OR ASSOCIATED ROWS IN A TABLE

User wants to find all observations done with a particular instrument configuration within a period of time where *e.g.* the CCD temperature variation from one exposure to the next surpasses a given value.

User wants to find all nights within the database of seeing measurements during which the seeing value was *e.g.* less than 0.5 for longer than 3 hours.

SCENARIO 6: MATCHING IMAGES AND CATALOG INFORMATION

A) Cross-identification of sources from images. User wants to find catalog objects that match sources on a given image. The tool displays the image which is assumed to be astrometrically calibrated and accepts via mouse buttons input of coordinate pairs. Input of pointings list from disk is also supported. The tool queries a catalog server for objects found in the vicinity of the given pointings. The image display is overlayed with the relevant catalog sources allowing for brighter stars to be drawn as larger circles and with a choice of options for colour and scale.

B) Target acquisition preparation. User selects field of interest which is then extracted from the digitized sky survey and displayed. Catalog and/or target list information can be overlayed graphically upon request with a suitable choice of colors and/or symbols. All functions available in scenario 2 (guide star selection) are also available here. A list of selected targets can be saved on disk (local server).

7–6   Conclusions

The Archive Facility for the ESO Very Large Telescope will be built gradually in the course of the next years. Profitting from the experience gained in years of archive operations for the ESO NTT, the VLT science archive promises to be of high quality and high scientific return because:

▷ traceable science operations will make the information contents reliable;

▷ a mature data interface definition will allow data to be understood and remain useful in the long-term in spite of evolving instrumentation;

▷ a close coupling between the archive, the data reduction system and observation preparation tools will support maximum utilization of the observing facilities;

▷ data management tools will ease the handling and organization of large data harvests;

▷ an engineering archive will support close monitoring of instrumental performance and serve as a tool to analyze and improve overall efficiency;

▷ new–generation catalog and database browsers will enhance the usefulness of archive data through making it easily accessible.

Acknowledgments

The following persons have participated directly or indirectly in the definition of the VLT Archive Facility: E. Allaert, D. Baade, P. Benvenuti, P. Grosbøl, O. Hainaut, B. Leibundgut, B. F. Rasmusen, G. Ruprecht, J. Schwarz, J. Spyromilio, W. Zeilinger, A. Zijlstra. Their contribution has been very much appreciated.

The collaboration with colleagues from the CADC has been very productive and appreciated.

References

[1] Catalog/Database Facility Requirements for the VLT and Gemini Projects, 0.2, 28 Nov 94, OSDH-SPE-ESO-000000-0002
[2] ESO VLT Science Operations Plan, 0.91, 20 Oct 94, VLT-PLA-ESO-10000-0441
[3] *Implementation of the Flexible Image Transport System (FITS)* (1992): NASA Office of Standards and Technology (NOST), 100–0.3b

[4] Albrecht, M A, Albrecht, R, Adorf, H–M, Hook, R, Jenkner, H, Murtagh, F, Pirenne, B, Rasmussen, B F (1994): in *Handling & Archiving Data from Ground–based Telescopes* Albrecht & Pasian (Eds.), ESO Conf. and Workshop Proc. No. 50.

[5] Albrecht, M A & Benvenuti, P (1994): in *Handling & Archiving Data from Ground–based Telescopes* Albrecht & Pasian (Eds.), ESO Conf. and Workshop Proc. No. 50.

[6] Albrecht, M A & Grosbøl, P (1992): in *Astronomy from Large Databases II*, Heck & Murtagh (Eds.), ESO Conf. and Workshop Proc. No. 43.

[7] D. Baade, E. Giraud, Ph. Gitton, P. Glaves, D. Gojak, G. Mathys, R. Rojas, J. Storm, A. Wallander (1994): The Messenger **75**.

[8] Péron, M, Albrecht, M A, Grosbøl, P (1994): in *Handling & Archiving Data from Ground–based Telescopes* Albrecht & Pasian (Eds.), ESO Conf. and Workshop Proc. No. 50.

[9] Péron, M, Baade, D, Grosbøl, P, Albrecht, M A (1995): in *Astronomical Data Analysis Software and Systems IV* Hanisch, Brissenden & Barnes (Eds.), ASP Conference Series, *in preparation*.

[10] Pirenne, B, Albrecht, M A, Durand, D, Gaudet, S (1992): in *em Astronomy from Large Databases II*, Heck & Murtagh (Eds.), ESO Conf. and Workshop Proc. No. 43.

Access Pointers

[→1] ESO Homepage: `http://http.hq.eso.org/eso-homepage.html`
[→2] ESO Archive: `http://arch-http.hq.eso.org/ESO-ECF-Archive.html`

8

Deep Near Infrared Survey of the Southern Sky (DENIS)

Erik Deul, Nicolas Epchtein & Jean Borsenberger

8–1   Context

1.1   BACKGROUND

Some 18 European and South American Institutes collaborate in the DEep Near Infrared Survey of the Southern Sky (DENIS) to provide a full digitization of the southern sky at wavelengths, similar to the Two Micron Sky Survey (TMSS). That survey was carried out some 25 years ago and has a sensitivity limit 10 magnitudes brighter than DENIS. The DENIS survey started end 1994, and the expected duration is at least four years. ESO has granted this project the Key Program status, and provided full allocation of the telescope during the planned data acquisition period.

The ultimate goal of this project is to provide the astronomical community with a complete digitized infrared map of the southern sky and a catalog of extracted objects, both of the best quality. Easy access to such data is essential. This is achieved through distribution centers, dedicated software packages, specialized catalogs, and assistance from the data analysis center personnel.

There are two data analysis centers to handle the shear data quantity and associated complexity of reduction. Their roles are complementary. The data analysis centers are located at the Leiden Observatory, the Netherlands, and the Institut d'Astrophysique de Paris, France.

1.2   DESCRIPTION OF THE INSTRUMENT

The 1m photometric telescope of ESO, at La Silla, is used to fully map the southern celestial hemisphere simultaneously in three bands: J ($1.25\mu m$), K' ($2.2\mu m$) and I ($0.8\mu m$) with CCD and NICMOS array detectors. The spatial resolutions are 3", 3", 1.0" and the limiting magnitudes are 16^m, 14.5^m, 18^m,

D. Egret and M. A. Albrecht (eds.), Information & On-Line Data in Astronomy, 73–85.

respectively. The observations are performed in a step-and-stare mode, using 'strips' consisting of 180 overlapping frames, 12' (one frame) wide in right ascention, and 30° long in declination. For telescope pointing and brightness calibration purposes additional frames accompany each unit. DENIS yields a huge quantity of data - one million survey fields will be observed, which amounts to several Terabytes of data.

The Paris Data Analysis Center (PDAC) is responsible for archiving and preprocessing the raw data to provide a homogeneous set of data suitable for both the Leiden and Paris data analysis streams. The Leiden Data Analysis Center (LDAC) extracts objects, ranging from point sources to small extended sources, parameterizes, and archives them into a source catalog. The LDAC also performs the photometric and astrometric calibration. For those sources flagged as extended by the LDAC, the PDAC extracts and archives images thus creating a catalog of galaxies. In addition the PDAC will provide means to mosaic the individual images.

8–2   Description of the databases

The ultimate goal of the DENIS Survey is to provide the astronomical community with a set of readily usable databases: a complete digitized 3–color infrared map of the southern sky, and a series of catalogues and databanks of extracted objects. An easy and fast access to the processed data will guarantee the best astronomical exploitation, and the Data Analysis Centers strive to reach this objective.

The final DENIS products are:

1. A Bright Source Catalogue extracted in real–time at La Silla for preliminary investigation and to prompt objects of astronomical interest for fast follow–up, later re-processed at LDAC. It will be distributed in printed and machine readable form.

2. Small Object Database that can be remotely interrogated. The LDAC is in charge of the database and the access software tools. A condensed form of this database, the Small Sources Catalog will be available as a catalog on CD-ROMs, to be updated and distributed regularly.

3. Extended Sources catalog in the form of an image database and catalog (of galaxies) that can be remotely interrogated. The PDAC is in charge of implementing this database and its access software. The catalog will contain cross-identifications with galaxies as recorded in the Lyon-Meudon Extragalactic Database (LEDA) (see also chapter in this volume, page 115).

4. A Processed Frame DAtabase of the full surveyed sky, also remotely

accessible. Tools to do mosaicing are provided by PDAC to allow creation of 'big' images of any part of the sky.

5. The Calibration star database (CALIB*) contains all standard stars that are used for the photometric calibration of the survey. It is implemented in Chile and Leiden.

During the data acquisition period, however, a number of working databases will be maintained. These databases store the raw information, but through dedicated interfaces users will be able to extract calibrated astronomical information in the 'final product' form. For more details see section 5–4 below.

8–3   Scientific contents of the database

The 0.8 to 2.2μm spectral range corresponds to the peak-emission of different species of astronomical objects. The DENIS survey will provide information on M dwarfs and subdwarfs, evolved giants, AGB stars, young stellar objects and brown dwarfs. The number of galaxies observed will be probably more than one million. The normal galaxies in the sample will have $z = 0.05$ to 0.2, and therefore will not be affected by evolution effects.

One major impact of this survey will be to provide digitized maps in a newly explored spectral range, that will facilitate stellar and galaxy counts followed-up by statistical investigations. Thanks to a better spatial resolution and a much lower sensitivity to cool dust emission, DENIS will be less limited by confusion in crowded regions (galactic plane, bulge, *etc.*) than was the IRAS (12μm) survey. The moderate interstellar extinction in the near-IR range compared to the optical range ($A_K \sim \frac{A_V}{10}$) will allow probing of stellar populations and of external galaxies at low galactic latitude and in optically obscured regions of our Galaxy.

8–4   Organization of the data

The duration of the project and the data dissemination period makes it necessary to allow user access to the data before the date of the last strip acquisition. The DENIS team has introduced the notion of working databases and final catalogs. Because during most part of the data acquisition period the general user will have to interrogate the working databases, we start with their description.

4.1   WORKING DATABASES

During the data acquisition period both data analysis centers will keep working databases that store the information particular to their reduction tasks.

These working databases form the principal source of information for the DENIS final products. After a delay period, in which quality checks are performed, the data are released for public use. Upon extraction the 'raw' data from the working databases will be converted to adhere to the 'final product' format. Thus the general user will interrogate the working databases to obtain DENIS data, but upon retrieval will get 'final product' information.

4.1.1 The Leiden Data Analysis Center (LDAC) The LDAC concentrates on the extraction of "small objects" from the DENIS images at all three proposed wavelength bands. Production of a catalog is done during the survey data acquisition and results in a first order data product within approximately one year after the observation time.

The main database at LDAC will be the Small Objects DAtabase (SODA). It stores the principal derived parameters for each extracted object as well as the conversion algorithms and parameters to derive astronomically meaningful units. During extraction the user may merge the individual entries of a multiple observed source into a single source parameter list. Note the nomenclature: an object is a record in the SODA database, it is the set of information derived for a successful object extraction from a frame triplet (I, J, and K bands). A source is is an astronomical object described by one object record or the set of information obtained after combining, using a well defined algorithm, the information of individual object records that store data on the same astronomical source.

All intensity enhancements in the maps that adhere to a set of criteria are extracted as objects (in general terms sources above the 5σ level). These objects are deblended and their positions calculated using the frame centers computed with the help of cross-identifications with Guide Star Catalog objects. To obtain better positions, strip-wide astrometric solutions are computed. The small objects are photometrically calibrated using a set of DENIS standards distributed around the southern sky. As this is an ongoing process the quality of the photometry progresses with time.

Basic information stored in the catalog consists of astrometric position, photometric intensity, image classification, and geometry. These parameters are derived using the moments, up to the second order, of the objects pixel distribution. Other parameters like pixel range over which the objects are detected, blend flag denoting the membership of a deblended object, image quality, and comparable ones are also saved.

The database is perceived by the user as one table, containing a mix of the above mentioned attributes. The individual entries in this hypothetical database are those of the Objects, but then in astronomical units. These units are determined on extraction using the most recent and up-to-date versions of the calibration parameters.

The SODA database is designed in accordance with the relational database model in mind, and is at the time of writing, still under development. It contains 4 entities, survey, strip, frame and object.

Survey is the basic database view of the user, *Strip* encompasses all information pertaining to the standard unit of measure in the DENIS survey mode. It includes information on calibration observations, position determinations, survey frames, *etc.*, *Frame* describes the individual frames and stores all information to be derived from or to be assigned to a single frame. The *Object* entity is the container of all information related to a single object extraction (all passbands).

Each of these entities has an attributes list attached . We give here a sketch of the attributes, not claiming to be exhaustive.

SURVEY Although this entity contains no individual attributes, it maps, through the relations the attributes from other entities. Therefore a SODA user accessing the Survey view sees *e.g.* the RA & DEC of a source because such parameters are derived from the intrinsic database attributes.

STRIP This entity stores information related to a strip, which is the DENIS unit of measure, e.g. the RA center of the strip, start point of the strip in Dec, the date of observation, etc.

FRAME The frame information stored in these attributes encompass such things as (Ra, Dec) frame offsets, exposure time, grey filter usage, type of frame (calibration, position, or survey), *etc.*

OBJECT The attributes for the object entity contain the raw extraction parameters. Some of these parameters are (similar for all bands): x-pos, y-pos, 2nd order moments, Flux, and several flags like border, single/blended, saturated, extended, near glitch, detection flags. Local image information of the type average background level and *rms* is also stored.

Among the individual entities in the SODA database there are several relations. These relations originate from the data acquisition method and the source extraction algorithm. For the SODA the following relations have been defined (not an exhaustive list):

SURVEY–STRIP The survey is built from a list of survey and non-survey strips, e.g. additional calibration measurements.

STRIP–FRAME Strips in the survey usually contain ≈ 180 frames, being either normal survey frames, or calibration or position frames.

FRAME–OBJECT The list of extracted objects is stored in this relational table. Frames can be either survey frames or calibration frames, with the result that the objects will either be survey or calibration objects.

OBJECT–OBJECT This particular relation stores the information that is obtained while performing a deblending extraction. It allows referencing to all objects from a single blend. In addition this relation stores the links between the individual extracted object entities from neighboring frames. These frames may either be in the same strip or from two neighboring strips.

OBJECT–ICAT For the extracted objects that have been associated with ICAT (Input CATalog = HST Guide Star Catalog) sources this relational table stores this information. This allows later re-processing of the astrometry using an ICAT with higher precision.

4.1.2 The Paris Data Analysis Center (PDAC) The PDAC will manage two major tasks in the processing of the data flow. The first task is to receive and clean–up the raw data, archive them locally, and forward a copy of the *cleaned* data to the LDAC. The raw data will be kept on DAT tapes at the PDAC. Information on the whereabouts of the data and the main observational parameters will be stored in a relational database. This also provides the means to retrieve the actual images from the data frame store. The second task of PDAC is to extract extended source information and make that data available to the astronomical community. This final PDAC product, the Extended Source Catalog will contain cross- identifications and source classification as derived from the Lyon Extragalactic Database (LEDA).

DENIS yields some 3–million elementary monochromatic images, arranged in more than 15 000 strips as input. Although this first organization is well suited to data reduction tasks such as flat–fielding, it is no longer a good packing for looking at compact areas on the sky, or at areas that are located within overlapping strips.

The data, while being processed, are reorganized. The first step will be to put together the three *images* of the same *frame*. This will be done prior to archiving at PDAC and shipping to LDAC. The second step will be to reorganize the strips to produce square groups of images. This will be done for the long–term storage of the data (probably CD-ROMs) during the final phase of the experiment. This latter organization on direct–access mountable media will provide easy access to three channel image data.

The Follow–Up Relational Base for Images (FOURBI)

To rapidly locate data, keep track of the main experimental parameters, and maintain links between field images and their calibration images the PDAC has developed a relational database under ORACLE, the *Follow–Up Relational Base For Images* (FOURBI).

This database will keep track of all external media on which an image is stored. It will also hold the main parameters related to images, *i.e.* PSF coefficients, and experimental parameters (instrument and sky). The FOURBI database contains 5 entities: tape, strip, slot, field image and calibration image. *Tape* contains the information relevant to the external storage media, *Strip* is the basic unit of measure in DENIS and stores data associated with this unit. *Slot* represents the sky location of the strips and is used for administrative purposes. *Field Image* holds information specific to a single frame, and *Calibration Image* is used for the frames measured for calibration purposes, both astrometric and photometric.

Each of these entities stores additional information in the form of an attributes list. We give here a sketch of the attributes.

TAPE has attributes to identify the tapes by name and number. In addition the two types of tapes, mono-chromatic and color tapes are marked.

STRIP stores such information as strip number, observer name, and date of observation.

SLOT The sky has been divided into a number of slots, the number is stored as an attribute.

FIELD IMAGE Information stored in this attribute list are items associated with the image sky position, telescope and environment characteristics, and several lists of calibration specific parameters.

CALIBRATION IMAGE keeps information on the type of calibration image.

Several relations among the FOURBI entities have been defined. These relations originate directly from the acquisition methods and the image processing scheme. For the FOURBI the following relations have been defined:

TAPE–STRIP Strips of images are store on DAT tapes either in raw mono–chrome, or three color (band) format.

STRIP–SLOT Each observed strip falls in one of the defined sky slots.

STRIP–FIELD IMAGE A strip contains a set of (usually 180) frames of survey images.

STRIP–CALIBRATION IMAGE Before and after a strip calibration frames are taken that will be used to calibrate the field images from that strip.

FIELD IMAGE–CALIBRATION IMAGE Each frame in a strip will be processed using calibration information obtained from the strip–associated calibration images.

This database is crucial for the experiment book–keeping. For this reason, all programs that work on, or handle the data, access it through a suitable interface.

The Processed Frame Database (PROFDA)

The PROFDA is the database in which all images are archived. It is, by far, the largest of the DENIS databases. The FOURBI is the main catalog through which access to the PROFDA (pixel information) can be obtained. As the image data is moved to an automatic mounting system, the FOURBI interface will be extended to actually allow retrieval and mosaicing of the image data. The image data is stored as FITS files.

The Large Object DAtabase (LODA) The PDAC deals with the large source extraction, storage and analysis. The extracted sources will be put in a standard non–relational database, called the Large Object DAtabase–base (LODA).

The LODA is built in a standard non-relational database scheme, using the multiple indexing mechanism. The objects are identified by the image slot number on the sky, and x, y position in the currently "best" image (at this time only one image is used, objects from lower quality images are kept in another database). Each object in the LODA constitutes a record composed of several attributes:

ASTROMETRY Using the available astrometry, the sky coordinates will be computed and indexed as they are the most commonly used keys to access the data.

PHOTOMETRY The main file will contain all extraction parameters, and magnitudes determined using available photometry. The way the fluxes will be calibrated is likely to evolve with time, until the ultimate extraction procedure is fixed.

OVERLAPS A system of cyclic pointers is maintained to link sources identified in several frames. On request the user may choose to get all of the chained sources, or only the "best" one.

RELATIONS TO OTHER DATABASES Records in the database contain pointers to a secondary smaller image database that holds the subimages of extracted sources, including the immediate neighborhood. Associations with existing catalogs will be made for known sources. The association process is handled by the Lyon Observatory.

4.2   THE FINAL PRODUCTS

After the last strip has been observed and the associated frame data processing has been completed final catalogs will be produced. These catalogs will contain the most accurate astrometry and photometry available, and are the result of the last processing step in the reduction chain.

4.2.1   The Infrared Bright Source Catalogue (IRBSC)   This database is aimed at providing the fastest access possible to the most prominent near–IR sources that could have escaped previous surveys at other wavelengths and at prompting immediate follow–up observations. Being extracted in real time, positions and photometry of these objects will not be highly accurate. At

Leiden, after a some time delay, more accurate positional and photometric information will be available through extractions from the SODA.

At the end of the survey, it is planned to produce a printed reference catalogue of bright 2–micron objects of a size comparable to the Bright Star Catalogue.

4.2.2 The Small Source Catalogue This catalog is compiled from the extracted objects as stored in the Small Objects DAtabase. It will contain some 10^8 sources of which 10^6 are expected to be galaxies.

Basic information to be stored in this catalogue consists of astrometric position (RA & DEC), photometric intensity (magnitudes), image classification parameters, and geometric parameters.

The SODA stores the astrometric information as obtained from the intensity weighted pixel sums; photometric information from the intensity sum, and pixel counts together with saved values for background and noise levels. The image classification basically separates stellar objects from extended sources. The geometric information such as semi–major and –minor axes and position angle are derived from the SODA second order intensity weighted pixel sums. Other parameters like pixel range over which the objects are detected, blend flag denoting the membership of a deblended object, image quality, and comparable ones will not be saved.

Association of extracted objects with other astronomical catalogues will be done purely on positional agreement, no attempt on associating objects using energy distribution characteristics or other astronomical parameters will be done.

The Small Source Catalogue will become available in electronic form either through remote access at the Data Analysis Centers or by means of a number of CD-ROM disks.

4.2.3 The Extended Source Database (ESDB) The ESDB will contain about 10^6 objects and will be extracted from the LODA. It will make use of a set of pointers included in LODA to provide the functionalities of a catalogue. The query procedure will allow redundant objects in overlapping fields to be merged.

This database will contain cross–identifications of known galaxies and their attached astrophysical parameters (*e.g.* radial velocity, HI data, UBV photometry, *etc.*) when DENIS suspected galaxies are cross-correlated with the Lyon Extragalactic Database (LEDA). DENIS images of these galaxies and their environment will also be stored in LEDA.

4.2.4 The Processed Frame Database (PROFDA) The PROFDA is both a working database and a final database. The final version will contain a collection of several 10^6 FITS images with extensive headers describing both the environmental conditions at the time of observation as well as the data processing steps they have gone trough. Depending on the availability of automatic mounting of the recording media, full and quick–response access will be available using the FOURBI as a search tool.

4.2.5   The calibration star catalogue (CALIB*)   The Calibration star database (CALIB*) contains all stars used in the DENIS survey for photometric calibration. The final version of this database contains not only the primary standard stars and etalon stars , but will also contain information on those objects used as secondary standard stars. The secondary standard stars are those objects that have proven to be stable photometrically throughout the survey period. They are also used in the final calibration procedure. Information stored in this database are the positional and DENIS photometric parameters for each star.

8–5   Access mode

5.1   THE SMALL OBJECTS DATABASE (SODA)

5.1.1   Availability   The SODA database will be on line at LDAC from the start of the data acquisition. External users will have access to information from this database after a one year period in which the data are being certified. There are four ways to interrogate the SODA database.

MOSAIC INTERFACE Through the World-Wide Web, via *e.g.* a Mosaic interface, the database can directly be interrogated. The menu driven system is provided through the Leiden WWW service [→1].

REMOTE TERMINAL Through a telnet session the user gets an instantiation of the SODA database query manager. He can create his personal databases as subsets of the full database by placing the appropriate SQL like queries. These personal databases can be exported to standard output formats (*e.g.* FITS binary tables).

LOCAL ACCESS The potential users may travel to Leiden and use the LDAC equipment to perform the actual queries and data extraction. In addition to the services as described in the previous items, a variety of added software packages are at their disposal to do initial astronomical data analysis work.

MAIL QUERY The user sends a mail message to the LDAC database server, that will perform the required operation on the SODA.

The result from the queries can be sent to the user either through regular mail, DAT tapes, or if data size allows through anonymous ftp. In all cases the LDAC personnel will assist the user at any stage of data retrieval when necessary.

5.2   THE LARGE OBJECTS DATABASE (LODA)

5.2.1   Availability   The LODA database will be on line at PDAC from the start of the data acquisition. External users will have access to information from this database after a one year period in which the data are being certified. In principle there are four ways to interrogate LODA.

MOSAIC INTERFACE   A World Wide Web server has been installed at PDAC to provide access to the databases [→2]. The interface is intuitive and provides a standard means of querying the LODA. The provided set of capabilities is somewhat limited compared to the remote terminal access method.

REMOTE TERMINAL   A telnet session provides a full screen menu–driven interface from which an ASCII output (screen or file), or a personal database can been made. The query process can be iterated on personal databases that have the same structure as the LODA.

LOCAL ACCESS   For large and complex interrogations a documented programmatic interface to LODA is available. PDAC staff will assist the user who will have full user access to the system supporting LODA (excluding writing to LODA itself).

MAIL QUERY   The user sends a mail message to the PDAC database server, that will perform the required operation on LODA.

The result from the queries will be made available via anonymous ftp. The user is advised through e–mail of the name and lifetime of their data on the FTP server.

5.3   THE PROFDA/FOURBI

5.3.1   Availability   As soon as an automatic mounter (jukebox) is available, the PROFDA can provide the image data to a permanently mounted direct–access device. In the mean time limited access will be provided. Queries to the image database are done through the FOURBI interface.

MOSAIC INTERFACE   The same World Wide Web server providing access to the LODA serves the FOURBI databases interface. With a uniform interface, comparable to the other DENIS query panels, the FOURBI queries can be composed.

REMOTE TERMINAL   A full screen menu–driven interface can be used through remote login. The interface has the same functionality as the Mosaic one.

MAIL QUERY The user sends a mail message to the FOURBI database server, that will perform the actual queries.

The result from the queries are image files on local PDAC disks or on DAT tape. Image data can also be made available through anonymous ftp. The user is informed by e–mail of the availability and state of his data.

5.4   THE CALIB*

5.4.1   Availability   This simple database will be made available through the same means as the SODA database, or by providing an ASCII copy of the individual table entries.

8–6   Future plans

On a long term plan, we will re-process all of the data making use of the expertise gained through the in-depth analysis of small pieces of the survey. The efforts developed to improve selected images will naturally build up the knowledge and tools necessary to re-process all of the data in a systematic and consistent way. It is likely that the on–site processing will not be homogeneous over the whole survey because, while the observations go on, as we learn about the data, we will find ways of improving the reduction algorithms. Because it is scientifically an important requirement that the end–product of the survey is processed homogeneously, we must be careful not to start such systematic processing before all problems with the data and reduction software and possible solutions have been well explored. It is clear that the data–reduction teams will need to be in close contact with scientists analyzing the data to progress in this direction.

It is also our aim to eventually combine the DENIS southern data with the northern data that the 2MASS project will produce. The final processing will thus definitively need to be done in close collaboration with the American team.

Access Pointers

[→1]   DENIS: http://eems.strw.leidenuniv.nl/denis/index.html

References

[1]   N. Epchtein (ed.), A Deep Near Infrared Survey of the Southern Sky (DENIS).
[2]   B.M. Lasker *et al.* , (1990) Astron. Journal **99**, 2019.

[3] G. Neugebauer & R.B. Leighton, (1969) Two Micron Sky Survey, NASA SP-3047.

Access Pointers

[→1] Leiden Data Analysis Center: `http://www.strw.leidenuniv.nl/denis/`
[→2] Paris Data Analysis Center: `http://denisexg.obspm.fr/`

9

Data from the Cosmic Background Explorer (COBE)

David Leisawitz & John C. Mather

9-1   The COBE Mission

The COBE[1] satellite was developed by the NASA Goddard Space Flight Center to measure the diffuse infrared and microwave radiation from the early universe, to the limits set by our astrophysical environment. It was launched on November 18,1989 and carried three instruments, a Far Infrared Absolute Spectrophotometer (FIRAS) to compare the spectrum of the cosmic microwave background radiation with a precise blackbody, a Differential Microwave Radiometer (DMR) to map the cosmic radiation precisely, and a Diffuse Infrared Background Experiment (DIRBE) to search for the cosmic infrared background radiation. The cosmic microwave background spectrum was measured with a precision of 0.03% (Mather *et al.* , 1994), the spectrum of the cosmic dipole was measured (Fixsen *et al.* , 1994), the background was found to have intrinsic anisotropy for the first time, at a level of a part in 10^5 (Smoot *et al.* , 1992 and Bennett *et al.* , 1994a), and absolute sky brightness maps from $1.25\,\mu m$ to $240\,\mu m$ have been obtained to carry out the search for the cosmic infrared background (Hauser *et al.* , 1991). As planned, COBE ceased collecting science data on December 23, 1993. The instruments that required cryogenic cooling (DIRBE, at wavelengths longward of $3.5\,\mu m$, and FIRAS) previously had stopped operating when the supply of liquid helium was exhausted on 21 September 1990. A more complete description of COBE is given elsewhere (Boggess *et al.* , 1992).

In addition to the mission's primary scientific objectives, which are cosmological, the data from all three COBE instruments can be used to study celestial sources of emission ranging from dust in the solar system to dust, gas, and stars in the Milky Way and a few nearby external galaxies (see references [7] to [14]).

---

[1]COBE is supported by the NASA Astrophysics Division.

D. Egret and M. A. Albrecht (eds.), Information & On-Line Data in Astronomy, 87–93.
© *1995 Kluwer Academic Publishers.*

COBE was designed, built, integrated, and tested at the NASA GSFC with scientific guidance from the Science Working Group (SWG). The final data processing and analysis also take place at GSFC. The COBE SWG is responsible for the definition, integrity, and delivery of the public data products, and includes the following members:

NAME	AFFILIATION	SPECIAL ROLE
J. C. Mather	GSFC	Project Scientist and FIRAS Principal Investigator
M. G. Hauser	GSFC	DIRBE Principal Investigator
G. F. Smoot	UC Berkeley	DMR Principal Investigator
C. L. Bennett	GSFC	DMR Deputy PI
N. W. Boggess	GSFC–ret	Deputy Proj. Scientist
E. S. Cheng	GSFC	Deputy Proj. Scientist
E. Dwek	GSFC	
S. Gulkis	JPL	
M. A. Janssen	JPL	
T. Kelsall	GSFC	DIRBE Deputy PI
P. M. Lubin	UCSB	
S. S. Meyer	MIT	
S. H. Moseley	GSFC	
T. L. Murdock	Gen.Res.Corp.	
R. A. Shafer	GSFC	FIRAS Deputy PI
R. E. Silverberg	GSFC	
J. Vrtilek	NASA HQ	Program Scientist
R. Weiss	MIT	Chairman of SWG
D. T. Wilkinson	Princeton	
E. L. Wright	UCLA	Data Team Leader

9–2 COBE Data Products

2.1 THE INITIAL PRODUCTS

An initial set of COBE data products was released on July 22, 1993. The FIRAS data cover the Galactic plane to an absolute latitude $|b| \leq 15°$ and provide spectra obtained in the 20 to 95 cm^{-1} band at 0.8 cm^{-1} spectral and 7° spatial resolution. The DIRBE data cover the Galactic plane, at 0.°7 resolution, over $|b| \leq 15°$ for ℓ within 30° of the Galactic center and over $|b| \leq 10°$ elsewhere, in 10 photometric bands ranging in wavelength from 1.25 μm to 240 μm. The DMR data include all-sky maps from the first year of observations at 7°

resolution at each of three frequencies: 31.5, 53, and 90 GHz, as well as calibrated pixel-ordered data. The data are presented in FITS binary tables.

An Explanatory Supplement is available for each of the three COBE instruments. These documents describe the intruments, their calibration, the data processing, and the resulting data products.

2.2 FURTHER DATA RELEASES

A second data release (June 1994), includes all-sky, full spectral range DIRBE and FIRAS coverage, DIRBE polarimetry, and the first two years of DMR data. Time-ordered data from each of the three instruments will also be released then. The COBE *Proposer Information Package*, available online (see below), gives detailed descriptions of the data products. The Explanatory Supplements will be updated as new data products become available.

The remainder of the DMR data and the DIRBE data taken after the cryogen ran out will be released in 1995.

2.3 ADDITIONAL DATA SETS

If there is sufficient interest in the scientific community, and if funding permits, additional data sets, designed to enhance the utility of the data products described above, will be developed. Examples of such additional data sets are zodiacal emission subtracted DIRBE maps, a file containing a spectrum model and a residual spectrum for each pixel in the FIRAS map, and dipole- and Galaxy-subtracted DMR maps.

2.4 ORGANIZATION OF THE DATA

The time-ordered data are based on the COBE telemetry format, which has a major frame of 32 seconds containing 128 minor frames per major frame. Each record typically contains all the data for the major frame. The time-ordered data will be released in native format, which will be described in the instrument Explanatory Supplements.

Great attention has been paid to the coordinate systems, geometric projections, and storage structure for the map data. The ecliptic coordinate system (J2000) is the fundamental COBE coordinate system. The quadrilateralized spherical cube (Chan & O'Neill, 1975 and O'Neill & Laubscher, 1975), in which the celestial sphere is projected onto the six faces of a cube using a modified tangent plane projection, was chosen. When a cube face is divided into equally spaced rows and columns of pixels, each pixel is equal in area on the sky to every other pixel, and thus the maps preserve photometric integrity. The lines of latitude and longitude on a cube face are curved but correspond to only modest distortion of the shapes of objects. Each of the ecliptic poles lies at the center of a cube face. The DMR and FIRAS use cube faces of 32×32 pixels, while DIRBE has 64 times that many (256×256). The COBE sky maps are not rasterized but are stored instead in FITS binary tables in order of pixel number according to a quad-tree structure. In this nearest-neighbor, hierarchical scheme, for example, all 64 DIRBE pixels contained in a FIRAS/DMR pixel are numbered consecutively and stored contiguously.

Publicly available COBE analysis software (see below) can be used to make Aitoff and Mollweide projections, and to cast the data into traditional FITS image (*i.e.* raster) format.

2.5   HOW TO OBTAIN COBE DATA

The data now available, and the associated documentation, may be obtained by anonymous `ftp` to `nssdca.gsfc.nasa.gov` (equivalent IP address 128.183.36.23). Please provide your e-mail address as the password. The file "`AAREADME.DOC`" in the directory "`cobe`" contains further instructions. A sample `ftp` session is given below:

```
$ ftp nssdca.gsfc.nasa.gov
Name: anonymous
Password: <your e-mail address>
FTP> cd cobe
FTP> get aareadme.doc
FTP> cd initial_products
FTP> cd dmr.dat
FTP> ls
FTP> binary
FTP> mget *.fits
FTP> quit
$
```

Alternatively, the data and documentation may be obtained on tape by request to the

Coordinated Request and User Support Office (CRUSO)
NASA Goddard Space Flight Center
Code 633.4
Greenbelt, MD 20771

phone: 301-286-6695
e-mail: `request@nssdca.gsfc.nasa.gov`

The COBE data are also accessible on the World Wide Web at [→1].

9–3   Users of COBE Data

During the six month interval following the release of the initial data products, approximately 400 people downloaded COBE files. Among them are funded and unfunded guest investigators, teachers, and other curious individuals. COBE data now reside on computers in Belgium, Canada, Denmark, Finland, France, Germany, Italy, Japan, The Netherlands, Poland, South Korea, the UK, and the US.

3.1   THE COBE GUEST INVESTIGATOR PROGRAM

Funding for COBE data analysis is provided by NASA Headquarters through the Astrophysics Data Program, for which Research Announcements have been issued annually. Fourteen proposals to analyze COBE data were approved in 1993, and 15 in 1994. A *Proposer Information Package*, available online in the ftp directory COBE/PROJECT_DATA_SETS, provides information essential to proposers.

A Guest Investigator Facility was established at NASA/Goddard to provide technical support to the approved investigators. COBE analysis software, coded in Interactive Data Language (IDL), may be used at the Facility or by remote login to Facility workstations, or may be downloaded and used at an investigator's home institution.

Investigators are requested to include the following statement in their publication acknowledgments: "The COBE datasets were developed by the NASA Goddard Space Flight Center under the guidance of the COBE Science Working Group and were provided by the NSSDC."

3.2   A NOTE TO EDUCATORS

Like other scientists, members of the COBE Team want the public to appreciate our rapidly growing understanding of the universe. Thus, while the COBE Team has been busy processing and analyzing data, publishing results, preparing data products for the scientific community to use, and assisting guest investigators, it has also found ways to serve a larger audience. An overview of the COBE mission, and other general information, can be obtained by ftp. The AAREADME.DOC file mentioned above will lead you to the files that contain this information. In response to numerous requests, some GIF and JPEG images were made available. The images, which can be found in the DIRBE/IMG and DMR/IMG subdirectories, are accompanied by captions given in ASCII text files. Be sure to use the binary command, as in the sample ftp session, before you transfer the files. Note that the images are useful as teaching aids but not as research tools; while they enable one to see important qualitative features of the data, the color scale is not linearly related to the measured sky brightness. More images (*e.g.* DIRBE all-sky maps and FIRAS emission line maps) will be made available as the data are released. Please don't hesitate to send comments or suggestions.

Contact Information

For assistance in the use of COBE data, contact David Leisawitz by phone (+1- 301-286-0807), FAX (+1-301-286-1771), electronic mail (`leisawitz@stars.gsfc.nasa` or surface mail at the address given above. COBE Information can also be found on the World-Wide-Web at [→1].

Please address requests for preprints or reprints of COBE Team publications to Susan Adams as follows:

> Ms. Susan Adams
> NASA Goddard Space Flight Center
> Code 685
> Greenbelt, MD 20771
>
> *phone:*   301-286-4257
> *FAX:*   301-286-1617
> *e-mail:*   `adams@stars.gsfc.nasa.gov`

References

[1] J. C. Mather *et al.* (1994), *Measurement of the Cosmic Microwave Background Spectrum by the COBE FIRAS*, Ap. J., **420**, 439.

[2] D. J. Fixsen *et al.* (1994), *Cosmic Microwave Background Dipole Spectrum Measured by the COBE FIRAS*, Ap. J., **420**, 445.

[3] G. F. Smoot *et al.* (1992), *Structure in the COBE Differential Microwave Radiometer First-Year Maps*, Ap. J. Letters, **396**, L1.

[4] C. L. Bennett *et al.* (1994a), *Cosmic Temperature Fluctuations from Two Years of COBE DMR Observations*, Ap. J., **436**, 423.

[5] M. G. Hauser *et al.* (1991), *The Diffuse Infrared Background: COBE and Other Observations*, in AIP Conference Proceedings No. 222, "After the First Three Minutes," edited by S. S. Holt, C. L. Bennett, and V. Trimble, p. 161.

[6] N. Boggess *et al.* (1992), *The COBE Mission: Its Design and Performance Two Years After Launch*, Ap. J., **397**, 420.

[7] C. L. Bennett *et al.* (1993), *Non-cosmological Signal Contributions to the COBE DMR Anisotropy Maps*, Ap. J. Letters, **414**, L77.

[8] H. T. Freudenreich *et al.* (1993), *DIRBE Evidence for a Warp in the Galaxy*, in AIP Conference Proceedings No. 278, "Back to the Galaxy," edited by S. S. Holt and F. Verter, p. 485.

[9] M. G. Hauser (1993), *COBE/DIRBE Observations of Infrared Emission from Stars and Dust*, in AIP Conference Proceedings No. 278, "Back to the Galaxy," edited by S. S. Holt and F. Verter, p. 201.

[10] T. Kelsall *et al.* (1993), *Investigation of the Zodiacal Light from 1 to 240 μm Using COBE DIRBE Data*, in proceedings of SPIE Conference on "Infrared Space-borne Remote Sensing" **2019**, 190.

[11] R. Arendt *et al.* (1994), *COBE DIRBE Observations of Galactic Reddening and Stellar Populations, Ap. J.*, **425**, L85.

[12] C. L. Bennett *et al.* (1994b), *Morphology of the Interstellar Cooling Lines Detected by COBE, Ap. J.*, **434**, 587.

[13] T. J. Sodroski *et al.* (1994), *Large-scale Characteristics of Interstellar Dust from COBE DIRBE Observations, Ap. J.*, **428**, 638.

[14] J. Weiland *et al.* (1994), *COBE DIRBE Observations of the Galactic Bulge, Ap. J.*, **425**, L81.

[15] F. K. Chan and E. M. O'Neill (1975), *Feasibility Study of a Quadrilateralized Spherical Cube Earth Data Base*, Computer Sciences Corporation, *EPRF Technical Report 2-75 (CSC)*, Prepared for the Environmental Prediction Research Facility, Monterey, California).

[16] E. M. O'Neill and R. E. Laubscher (1975), *Extended Studies of a Quadrilateralized Spherical Cube Earth Data Base*, Computer Sciences Corporation, *NEPRF Technical Report 3-76 (CSC)*, Prepared for the Naval Environmental Prediction Research Facility, Monterey, California).

Access Pointers

[→1] COBE: `http://www.gsfc.nasa.gov/astro/cobe/cobe_home.html`

10

The NASA/IPAC Extragalactic Database

G. Helou, B.F. Madore, M. Schmitz, X. Wu, H.G. Corwin, Jr.,
C. LaGue, J. Bennett, & H. Sun

10–1 Introduction

The NASA/IPAC Extragalactic Database (NED) is an electronic research tool which provides uniquely powerful access to published multi-wavelength data on extragalactic objects. It consists of a computer database with a broad range of published information, and a user interface which allows fast and flexible retrieval of the information via the Internet. It is accessible anonymously and free of charge to any account on the networks, and used primarily by researchers, students and librarians. NED is physically located at the Infrared Processing and Analysis Center (IPAC) on the Caltech campus in Pasadena, California (USA).

NED provides an object-based representation of the extragalactic sky, with accurately located sources and coherently presented data, and direct links from this information to the refereed literature. NED is built around astronomical sources. Its starting point is a detailed merger of extragalactic catalogs, rather than a collection of juxtaposed catalogs. This merger is constantly enriched by new identifications and by the addition of objects appearing in short lists throughout the literature. The resulting database is augmented with a systematic bibliography, with well-characterized measurements for each object, and with notes from catalogs and the literature. NED can be searched for objects in a variety of ways, including constraints on positions, redshifts and type of object, and will plot the distribution on the sky of objects retrieved. NED can be searched for references by citation or by author name, and will display abstracts of papers published since 1988, and of theses from English-language institutions.

In the four years since its public release, NED has been adopted by the extragalactic research community worldwide. The response from users has been extraordinary, both in the enthusiasm expressed and the frequency of consultation. As of the Fall of 1994, this frequency hovered around 6,000 sessions per month, with hundreds to thousands of additional batch-mode

D. Egret and M. A. Albrecht (eds.), Information & On-Line Data in Astronomy, 95–113.
© 1995 *Kluwer Academic Publishers.*

requests and server-mode queries per month.

10–2   Context

Lists of galaxies were being assembled as part of the early surveys for diffuse nebulae (Messier, Heschel, etc.) even before the term "extragalactic" had been coined. The advent of photographic sky surveys led to the first catalogs of galaxies with tens of thousands of entries (*e.g.,* the CGCG by Zwicky *et al.*). The number of galaxies in catalogs has grown exponentially with time, from about thirty at the time of Messier to the millions being generated today by plate-scanning machines like APM (Cambridge University) or APS (University of Minnesota). At the same time, specialized catalogs of galaxies (interacting, low surface brightness, *etc.*) and of quasars have also multiplied.

Over the past three decades, advances in technology finally made it possible to observe in spectral windows other than the visible range both from the ground and from space. The exploration through these new windows naturally turned into sky surveys, resulting in catalogs of radio, x-ray, and infrared sources. The fraction of extragalactic sources in these surveys have ranged from the substantial to the dominant.

All sky surveys have been instrumental in advancing astronomy. Catalogs are an invaluable tool to statistical astronomy, and a permanent source of surprises, of new classes of objects hiding among "normal" sources. Much is learned from the comparison of a new catalog to existing ones. For example, the comparison of IRAS Point Sources to optical catalogs of galaxies revealed ultra-luminous galaxies (Soifer *et al.* 1984); their comparison to radio surveys led to the infrared-radio correlation (Helou *et al.* 1985). However, with the accumulation of surveys and overlap in catalog membership, it has become exhorbitant to operate only in terms of catalogs, carrying $(N - 1)$ potential connections for each entry in each of N catalogs. In addition, this mode of operation impedes the efficient use of the information to gain physical insights, making it harder for instance to weigh evidence on positional coincidence against evidence on emission properties.

In parallel, the astronomical technical journals have been witnessing their own explosive growth in the publication rate. Abt (1988) estimates that the number of articles related to galaxies stood at 1,500 for the publication year 1985, and that it is doubling every eight years. Galaxies and cosmology are the fastest growing among eleven sectors of the astronomical literature surveyed by Abt.

The trend of increasing quantity and diversity of data will continue, driven by constantly improving astronomical instrumentation and data processing capabilities, and by large space missions such as the Hubble Space

Telescope (HST), the Infrared Space Observatory (ISO), the Advanced X-ray Astronomy Facility (AXAF), and the Space Infra-Red Telescope Facility (SIRTF), as well as ground-based large surveys (2MASS, Sloan, etc.). This trend makes it harder for individual researchers to stay up to date, in terms of being aware of new data, and tracking new ideas. This is the dual challenge posed by explosive data growth: dealing with the sheer volume, but also interconnecting intelligently the huge variety of information available.

In early responses to this challenge, attempts were made to consolidate the published results into reference catalogs (*e.g.*, the effort by de Vaucouleurs & de Vaucouleurs (1964) which culminated in the publication of RC3 in 1991), or more specialized compilations of data such as the Palumbo *et al.* (1983) Catalogue of Radial Velocities, the Huchtmeier and Richter (1989) General Catalog of HI Observations, or the Catalog of Infrared Observations (Gezari *et al.* 1993). In the early 1980's the Centre de Données Stellaires of the Observatoire de Strasbourg introduced a new approach to dealing with data, namely a computer database service called SIMBAD (Set of Identifications, Measurements and Bibliography for Astronomical Data), which focused then on stellar data. The innovation was in the exclusively electronic archiving and interrogation, with continuous updating of the contents. Since then, a variety of systems and services have been set up, making up collectively a significant element of the astronomy research environment, as this book amply demonstrates.

In 1987, a fortuitous confluence of interests and opportunities, and the inspiring success of SIMBAD, prompted a number of IPAC astronomers to propose to NASA to establish a new extragalactic database that would be kept up to date in bibliography and published data, and be open to the astronomical community. IPAC as an environment was well suited to such an enterprise in terms of infrastructure, expertise, and experience with catalogs. Work on the design and implementation of NED started in June 1988, and its public release was announced at the 176th Meeting of the American Astronomical Society at Albuquerque in June 1990. NED has been in operation continuously since then.

10–3 Scientific Contents

NED is an object-oriented database, meaning that it organizes all information around individual extragalactic objects as opposed to leaving this information stored in catalogs or compilations; catalog membership is but one of many object attributes. In database management terminology, NED is organized as a relational database. It is built around a set of tables called the Object Directory (3.1); this is the cornerstone of the database, to which additional tables are connected, usually via object identifications. These tables contain bibliographic references (3.3), published data of arbitrary description (3.4), notes from various sources (3.5), or copies of original catalogs (3.6). Such an architecture is easily extensible, so that other types of information (such as images or spectra) can be added in the future.

3.1 OBJECT DICTIONARY

The Object Directory is the master list of objects recognized by NED, along with their positions, redshifts, and basic attributes (3.2). It represents the systematic merger of some 40 major astronomical catalogs, as well as scores of shorter lists culled from the refereed literature, such as the Palomar-Green list of quasars (Schmidt and Green 1983), the Schombert and Bothun (1988) low surface brightness galaxies, or the Hickson (1989) compact groups. New lists and catalogs are continually folded in based on thorough cross-identifications for each object made by NED team members. As of this writing (November 1994), the Object Directory contains about 330,000 objects known by 650,000 names. Naturally, NGC and IC names and a few hundred special names (*e.g.*, Einstein Cross, Taffy Galaxy, etc.) have also been added to the Directory.

Table 10–1 shows some 20 catalogs with more than 4,500 members each, all of which have already been folded into the Object Directory. The abbreviations in the second column of Table 10–1 are standard nomenclature in NED, and are used in what follows to refer to these catalogs. Column 3 reports the number of objects from that catalog that appear in NED. This is often different from the count of entries in the catalog for two reasons: some of the catalogs include Galactic objects which are not folded into NED; and many catalogs include multiple objects appearing as single entries subsequently separated out by NED. The next set of challenges will be to fold into NED the large compilations of galaxies identified from automated scans of digitized optical sky surveys.

Most modern catalogs consist either of mostly Galactic or of mostly extragalactic sources (such as the PKSCAT90 radio catalog), or they will come with a relatively efficient prescription for distinguishing between the two populations (like the ESO/Uppsala or the *IRAS* Point Source Catalog). To appear in NED, an object has to have been classified as extragalactic in the literature, or to have a reasonable expectation (50%) of being extragalactic based on its observed properties. For instance, IRAS sources were selected for inclusion in NED only if (i) they had flux densities flat or rising with increasing wavelength, and (ii) were located in areas of the sky with low cirrus confusion noise (IRAS Explanatory Supplement 1987). Given such prescriptions, NED will unavoidably contain objects belonging to the Milky Way. We find their inclusion preferable to the exclusion of extragalactic sources; objects re-classified after having been mistakenly called extragalac-

Catalog Name	Abbreviation	Number of Entries
5 GHz Green Bank Survey	87GB	55,000
Galaxies in the Lick Proper Motion Catalog	NPM1G	50,517
IRAS Faint Source Catalog	IRAS F	49,011
1.4 GHz Green Bank Survey	[WB92]	30,239
Catalog of Galaxies and of Clusters of Galaxies	CGCG	29,418
Morphological Catalog of Galaxies	MCG	29,003
ESO/Uppsala Catalogue of Galaxies	ESO	18,438
ESO Surface Photometry Catalogue	ESO-LV	15,467
Third Bologna Radio Catalog	B3	13,353
Uppsala General Catalog of Galaxies	UGC	12,921
IRAS Point Source Catalog	IRAS	10,548
New General Catalog	NGC	7,021
Coma Cluster Catalog (Goodwin *et al.* 1983)	[GMP83]	6,726
Galaxies Behind the Milky Way (Saito *et al.* 1991)	CGMW	6,509
Arp-Madore Catalog of Peculiar Galaxies	AM	6,445
Kiso Ultraviolet Galaxies	KUG	6,348
Parkes Catalog 1990	PKS	5,991
Southern Galaxy Catalog	SGC	5,481
Abell Clusters of Galaxies	ABELL	5,250
Shapley 8 Cluster Catalogue (Metcalfe *et al.* 1994)	[MGP94]	4,904
Fourth Cambridge Radio Catalogue	4C	4,843
Flat Galaxy Catalog (Karachentsev 1993)	FGC	4,754
Hewitt & Burbidge Optical Catalog of Quasars	[HB89]	4,617

Table 10–1: Major Extragalactic Catalogs in the Object Directory

tic will remain in NED to signal that switch. On the other hand, we will also miss some extragalactic sources, a price to pay to avoid including many more Galactic objects; this omission is remedied whenever detailed work is published reporting the extragalactic nature of such objects.

For each object, the Directory contains positional data, names and "basic data" (3.2), and a "preferred object type", *i.e.* the most useful description of this object by one of the categories in Table 10–2 (*e.g.* "galaxy" or "quasar" is preferred to "infrared source" or "ultraviolet excess source").

Sub-galactic objects (HII regions, globular clusters, etc) within other galaxies are now infrequently found in NED, but may be included eventually in a systematic fashion. For these, we use object type abbreviations based on, and slightly simplified from, those used by SIMBAD (*e.g.* "SN" for supernova, "PN" for planetary nebula, *etc.*). The same object type abbreviations are used for objects within the Milky Way, except that they are prefixed by an exclamation mark to emphasize their Galactic location.

Much care goes into the collection of positions and redshifts into NED, and they are continually over-written by more accurate values as they become available. Positions are stored internally in the J2000 system, along with their uncertainty, and a reference to their origin. All published positional uncertainties are transformed to a representation as a 95% confidence ellipse, whose semi-major and semi-minor axes and position angle are kept. We also store Galactic coordinates, and total Galactic extinction in the blue at the position, derived from the Burstein-Heiles (1978) reddening maps. Ecliptic and super-galactic coordinates, and Equatorial coordinates other than J2000, are computed at display time.

3.2   NAMES AND BASIC DATA

The various names by which each object is known are stored and displayed in uniform NED formats, *e.g.* 4C +00.30 or UGC 00299, though users may use variations as a starting point for name searches. Each name is associated with an object type (see Table 10–2) which reflects the kind of survey that originated that name. Thus, even users not familiar with the specific surveys can see at a glance that UGC 12699 was recognized as a galaxy in eight different catalogs, and detected as an ultraviolet excess source (called MRK 0538), as an emission line source (UM 167), as an infrared source (IRAS 23336+0152), and as a radio source (87GB 233340.0+015229). The discovery methods associated with most entries in Table 10–2 are self-evident; an "absorption line source" is one revealed by absorption against a bright continuum source, typically a quasar; a "visible-light source" refers to an image with insufficient morphology for a definite classification as galaxy or star. "Other" sources are real objects so unusual or rare that they do not warrant a separate type

GClstr	Cluster of Galaxies
GGroup	Group of Galaxies
GTrpl	Triplet of Galaxies
GPair	Pair of Galaxies
G	Galaxy
QSO	Quasi-Stellar Object
RadioS	Radio Source
IrS	Infrared Source
VisS	Visible-Light Source
EmLS	Emission Line Source
AbLS	Absorption Line Source
UvES	Ultraviolet Excess Source
XrayS	X-ray Source
GammaS	Gamma-Ray Source
Other	Unusual object type
Q_Lens	Gravitationally Lensed Quasar
PofG	Part of Galaxy

Table 10–2: Extragalactic Object Types in NED

definition, such as isolated intergalactic clouds detected in emission.

NED uses for object names acronyms as suggested by the authors, or as in current usage. However, when an acronym is not provided for new objects, NED finds it necessary to create one to provide unique identification. In general, the acronym then used consists of the first initials of the first three authors' last names followed by the last two digits of the year of publication, with the whole string enclosed in square brackets; this is then followed by each object's identifier as given in the paper, or each object's coordinates as in the paper. Examples of such constructions are found in Table 10–1. NED also introduces suffixes to names of objects as required to resolve conflicting or overlapping identifications. For instance, a member of a pair identified only in the notes of the UGC would be referred to as UGC 08333 NOTES01; more commonly used are suffixes "NEDxx" when an ambiguity is resolved by NED, or "ID" when a radio or an infrared source is identified as a galaxy but no new name is provided.

"Basic Data" are the attributes most essential to a broad description of the object at hand. There is a different set of attributes for each object type, but each object has only one set of basic data, that corresponding to its "preferred object type". UGC 12699 for instance will only have basic data appropriate for a galaxy, namely an optical magnitude, a major and minor diameter, and a morphological description; On the other hand, a radio source has a flux density, the radio frequency of that measurement, a spectral index in the vicinity of that frequency, a size, and a morphology (*e.g.* head-tail); a quasar is described by an optical magnitude, and a qualifier (*e.g.* BL Lac or radio quiet).

These basic data are indicative values only, for they originate from many different sources not explicitly identified in the database. No attempt has been made to place them on a uniform scale. The main sources are catalogs and compilations, with the more accurate data sets favored, and the larger ones preferred at comparable accuracy. More rigorously defined and referenced data go into the photometry and positional data tables described in section 3.6.

Finally, "essential notes" generated by NED are attached to some objects (12,000 as of this writing) to point out significant facts, such as an erroneous identification, unique property, discordant value, or special relation to another object; these notes are always displayed along with basic data.

3.3   BIBLIOGRAPHIC REFERENCES

This segment of NED consists of pointers indicating the existence of useful information on a certain object in a given publication. A full bibliographic reference is kept for each publication in addition to a 19-character reference code which encapsulates the full reference in abbreviated form (see chapter 24 in this book).

The bibliographic references in NED derive from two main sources. Starting in 1988, members of the NED team have been reading systematically several of the major journals to identify papers presenting meaningful new information on extragalactic objects, as well as papers of extragalactic interest in general. This coverage is highly reliable and complete for *A&A, A&AL, A&AS, AJ, ApJ, ApJL, ApJS, MNRAS,* and *PASP* starting in 1990, for *IAU Circulars* starting in 1991, and for *Nature, PASJ, Astronomy Reports* and *Astronomy Letters* (the last two formerly *Soviet Astronomy* and its *Letters*) starting in 1992. This segment of the literature is now yielding about 75,000 pointers per year. The second source of bibliographic data is the SIMBAD project, which has kindly provided all of their references to extragalactic objects up to 1989, and has been providing updates on an annual basis since then. The SIMBAD pointers are complementary, since they are based on a search of many more astronomical journals conducted for many years by the librarians at the Institut d'Astrophysique de Paris, and which has produced, for extragalactic sources, systematic coverage starting in 1983, and sporadic coverage going back to 1917.

As of this writing, the NED database contains well over 500,000 pointers linking objects to over 25,000 distinct publications. Bibliographic pointers to the journals scanned by NED are typically available on-line about one month after the corresponding issues have appeared in print.

3.4 PUBLISHED MEASUREMENT DATA

One of the goals of NED has always been to collect and store information about new extragalactic data appearing in the literature. The goal was to carry fully those data that can be expressed in a few numbers, and have clear descriptions of those which cannot, such as spectra and maps. This ambitious long-term goal is being undertaken in steps, with the treatment of photometric data being the first major implementation to date.

Broad-band flux densities, fluxes and magnitudes at any wavelength, and fluxes in the 21 cm HI line and the 2.3 mm CO line are now routinely entered into the database, mostly from large compilations and catalogs, working towards an eventual systematic coverage of the literature. The effort concentrates exclusively on measurements of global or nearly global emission from objects. Each measurement is fully referenced and cast internally into a uniform "data frame" which includes the most significant information needed for a critical appraisal, *e.g.*:

▷ the as-published wavelength or frequency, flux value, units and uncertainty, or upper limit, and the meaning of the last two quantities, such as "2σ" or "plate limit";

▷ indications pertaining to the derivation in the spatial domain, such as "flux in fixed aperture", or "integrated from a map"; similarly for the frequency domain, such as "synthetic band" or "integrated over line";

▷ a variety of frequently recurring indications of a general nature, such as "from new raw data", "homogenized from previously published data", "extinction-corrected for Milky Way", or "K-correction applied".

NED currently offers about 635,000 photometric data frames covering the spectrum from the radio to the X-rays. As each measurement is integrated into NED, its equivalent value on a uniform system of units is also stored, making it possible to compare all data across objects and across the spectrum. Building on these coherent data, NED will soon offer its users a view of the broad-band spectral energy distribution using available data across the whole observable spectrum for any object.

Besides photometry, current plans call for similar data frames to be collected for position and redshift measurements, morphological classification and spectral or other typing.

3.5 NOTES

Almost every catalog published has a wealth of data appearing as notes on
individual objects, usually in an appendix that does not get circulated in
computer-legible form, and therefore remains largely untapped. Many journal articles also contribute valuable notes which are all too easily overlooked
or forgotten. NED has made a special effort to make notes available. In some
cases, this has entailed digitizing the printed material for the first time (*e.g.*
the Hubble Atlas of Galaxies (Sandage 1961), all 8,000 UGC notes (Nilson
1974)). In other cases, *e.g.* the ESO/Uppsala and the SGC, the shorthand
used in the notes was translated into regular English before the notes were
included in NED; in the case of the MCG or the Arakelian list (1975), the
contents of the notes were translated from Russian to English before entry.
As of this writing, the database contains over 34,000 individual notes. The
notes are retrievable by query on the object name, and are stored along with
a reference to their source, and the particular object name used by the author
of the note.

3.6 LITERAL CATALOGS

The major catalogs (such as those listed in Table 10–1 above) are the source
material from which NED assembles the Object Directory, and as such they
need to be accessible in their original form, especially if they contain substantially more information than NED has extracted from them. Under the title
of "literal catalogs", NED provides users with a view of the catalog entries
as close as possible to their original printed appearance. Currently available
are images of the RC3, UGC, ESO/Uppsala, PKS, and [HB89].

3.7 ABSTRACTS OF PAPERS AND THESES

In the course of scanning the core journals (3.3), papers of extragalactic
interest have their abstracts digitized by the NED team and integrated into
NED as text, and made immediately available to users for browsing. Besides
articles that contribute original data, the abstract collection includes reports
of theoretical studies, modeling, or empirical analysis. NED currently offers
about 10,000 abstracts on-line.

NED introduced in 1992 an on-line collection of dissertation abstracts of
extragalactic interest. Thesis abstracts (mostly from U.S. institutions) almost
complete back to 1980, and titles and authors dating back to 1909 were
generously provided by University Microfilms. New thesis abstracts are
immediately added to this collection once they have been accepted by the
granting institution and forwarded to NED.

10–4   Functions

The user's view of a database is defined by the data retrieval process. From this point of view, NED can be primarily described as a source of lists of objects and lists of bibliographic references, with several ancillary functions attached. There are many ways to construct lists of objects, which divide into two modes, local searches and global searches (4.1). Having retrieved a list of objects, members in that list can be singled out for follow-up, typically retrieving lists of relevant references (4.2), or of various types of data pertaining to these objects (4.3). Lists of references can also be constructed directly. It is also possible to branch in cascade from lists of objects to lists of references, although the interface will only have available the most recent list of each kind. Ancillary functions provided by NED include the plotting of the distribution on the sky of objects in lists, the browsing of paper and thesis abstracts, a coordinate conversion and precession utility, various information files (e.g. a list of upcoming conferences), session history tracking, and the transmission to the user by electronic mail of ASCII files containing data retrieved during a session.

Aside from these functions, there is an elaborate collection of software tools, largely invisible to the user, whose purpose is to populate NED with new objects and data and update existing information, while preserving data integrity and consistency, and maintaining traceability of the modifications.

4.1   SEARCHING FOR OBJECTS

Under the local search mode, users can locate objects by specifying a name, or a vicinity to search in. Substantial latitude is allowed in entering namesname resolver, with the input being interpreted by the NED interface and cast into the standardized formats used by NED for internal storage (section 5). Moreover, a search by name (*e.g.*, VV 136) will normally return (unless otherwise selected by the user) all objects that carry the specified name plus a suffix (*i.e.*, VV 136, VV 136a, VV 136b, and VV 136c). Search by vicinity returns all objects lying within a circle, whose radius and center are entered by the user. Such searches can be centered on an object specified by name, in which case the occurrences of suffixed versions of this name are ignored. If the center is specified by position, it can be given in any of four coordinate systems: Equatorial or Ecliptic (arbitrary equinox and epoch years), Galactic, or Super-Galactic. Searches by vicinity can include additional filters on object parameters, such as including or excluding object types, and restricting redshifts to certain intervals. In another local search mode, the user enters an abbreviation of coordinates in the "IAU style" commonly used to construct names from positions (*e.g.* 2254-054), and a positional search is made by NED covering the most likely area in the sky from which these coordinates might have originated.

The simplest global mode search retrieves all objects linked to a given reference. The most recent and powerful of the search modes allows users to search through the whole database by parameter, e.g. to find "all quasars at Galactic latitudes between 45° and 50° and with redshifts greater than one", or "all clusters of galaxies which are not known to be X-ray sources". Searches can now be made for objects satisfying simultaneously conditions on Equatorial and Galactic coordinates, redshift and object types; constraints on photometric data and other attributes will be added incrementally. These conditions are specified in terms of (i) coordinate intervals to include or exclude, (ii) of object types to include or exclude, and (iii) for redshifts (since not all objects have them) in terms of availability or lack thereof, as well as included or excluded intervals.

Each one of those searches returns a list of objects, along with the corresponding contents of the Object Directory and Basic Data (3.1 and 3.2 above), and with the number of bibliographic references (3.3), the number of photometry data points (3.4), and the number of notes (3.5) available for each object, thus presenting the user with a summary status of the object within NED. The interface displays this list, allowing the user to examine in detail individual objects and retrieve further information as detailed in the next section.

4.2   RETRIEVING DATA

Data retrieval requests presently in operation correspond to those data structures described in section 3 above. Users can list the bibliographic references for specific objects; once a reference has been found, its abstract can be displayed. Users can browse through the notes and the literal catalog entries linked to individual objects. Similarly, photometric data frames can be listed for each object, and examined in detail.

Requests for notes and references can be issued to follow up on an object identified in a previous search for objects, or can be formulated *ab initio* by specifying an object name. The second mode offers (as the default option) the advantage of an "extended" search, which starts by locating other objects which have a name root in common with the name specified, then retrieves all references to all objects thus related to the object of interest. For instance, the Hickson Compact Group HCG 23 consists of five galaxies called HCG 23A through HCG 23E. NED contains notes pertaining to, and references linked to, only some of the five individual members. A search for articles strictly relevant to HCG 23 will not reveal papers linked to individual member galaxies, whereas an extended search will return all references to HCG 23 as well as HCG 23A through HCG 23E. More useful yet, the same all-inclusive list will be returned by an extended search on NGC 1216 which also happens to be the galaxy HCG 23C.

10–5 Interfaces

The functions described above are provided via three distinct interfaces, all of which are served by the same underlying data and software. The direct human-machine interface (5.1) offers the most versatile and detailed mode of extracting information, and typically drives the design of new capabilities. However, most functions are also available in simpler forms in a batch mode via e-mail (5.2), and in a client-server mode for machine-to-machine interaction (5.3). There is no charge for logging into the service or retrieving data from it in any of the modes.

5.1 HUMAN-MACHINE INTERFACE

The user's view of a database is strongly colored by the ease with which one interacts with it. This ease is defined by several factors, including: (i) the amount of learning needed before sessions become productive; (ii) the degree of overlap between the specific questions a user has and those the database can answer; (iii) the power and flexibility available to the user in formulating a query; (iv) the convenience in the mechanics of submitting a search and collecting the results; and (v) the elapsed time between submitting a query and gettting the response. While some of these factors are determined by the internal structure of the database, they can be modified by, and in the end depend mostly on, the user interface to the database.

In line with those perceptions, NED's primary interface was designed as an interpreter that enables an astronomer or a librarian to use the database without learning its jargon, understanding its internal data organization, or being familiar with the interface mechanics. One of its main goals is to make a user's first session productive in terms of obtaining the sought-after data within a reasonable amount of time. This goal has been dubbed the "five minute rule": unless the interface yields useful data within five minutes of a new user's first session, that user might never return to the database.

The direct NED interface may be accessed by setting up a network connection to IPAC from any Internet host, using a command of the form <telnet ned.ipac.caltech.edu>. Once connected to the NED platform and prompted for a "login", the user should respond with "NED"; no password is needed. The interface then gives the user a choice between a character-based VT100 terminal mode or an X-Window-based graphical interface with plotting capabilities. These two modes of presentation run the same source code, compiled with different options, and access the same data. The user can then choose between them based on the hardware or software available at their end, and on the bandwidth of their connection to IPAC. The X-Window mode offers advantages in clarity of presentation and plotting abilities (sky distribution, spectral energy distribution), whereas the character-based mode requires a fraction of the network bandwidth to run at an acceptable pace.

Users control the NED session by selecting options and entering data within the context of a menu tree; the interface presents itself as a screenful of text for each node in this tree, with a standard screen format maintained throughout the session. This format consists of three areas: the one at the top displays available options, while the one at the bottom displays commands, and the middle box serves for input and output. Options are displayed as abbreviated descriptions of the functions they serve. As an option is selected (typically with a single key stroke or a cursor point-and-click), a new screen is displayed, a data input screen is activated, or a data output sequence is initiated. The commands perform simple, general purpose tasks such as terminating input or output, moving to a higher-level menu, obtaining detailed help, or ending the session.

A user would start a session, then select options until the desired input panel is reached; once the various fields in the selected panel have been filled out, a corresponding search is submitted, and the interface is inactive until the search concludes. For potentially time-consuming searches, the user may be warned and asked for verification before the search is submitted. The resulting data are then available for display in various output screens, until a new search is requested, causing the new results to overwrite the previous ones. Users can request all data returned during a session (including abstracts) to be sent to them by electronic mail.

An essential feature of the NED interface is self-documentation. In the first place, a NED session is steered at every step by a choice among options expressed in astronomical jargon, instead of a database interrogation language. Secondly, help functions are available at several levels throughout a session: (1) All menu items, options and input fields are labelled in simple terms, and every error message informs the user directly about what could not be processed. (2) A longer explanation of the functions and usages within every screen can be displayed by selecting the "HELP" option within that screen. (3) An introduction and overview, a general tutorial and news about recent additions to the system are also available as menu options in the first NED screen.

Similarly extensive documentation is available on output; data fields are labeled in whatever detail is necessary, and abbreviations are avoided as much as possible. On the screen displaying names of objects (cross-identifications) for instance, the user can ask for information on a given name, and get a brief description and a bibliographic reference to the catalog in which the name originates. In addition, users can tailor their output environment by selecting many of the display attributes, such as parameters for sorting lists, or coordinate system for displaying positions.

As the NED interface is accepting data from a user, it checks them to the extent possible for ambiguities, conflicts or errors; more significantly, it attempts to interpret them, and echoes its interpretation to the user for verification. These verification capabilities translate into less restrictive formats for the user to adhere to; for instance, the right ascension field will accept any of "6h0m5s", "6:0:5", or "6 0 5" as valid input, and interpret it as "06h00m05s".

The most versatile and sophisticated interpreter in NED is associated with names of objects, since naming conventions show great diversity across the literature. This interpreter is based on the "lex" regular expression processor; it gets the nomenclature guidelines out of a file, thus simplifying the task of updating name conventions recognized by NED. It usually generates one of four responses: the input is uniquely recognized, and rewritten in the standard NED internal format in preparation for querying the database; the input is ambiguous, so a number of possible interpretations are offered for the user to choose among; the input is distantly related to one or more naming conventions, so the proper formats for the latter are offered to the user as information; or the input is completely unrecognized by NED. Examples of each case are: "m1-2-3" is accepted as "MCG +01-02-03"; "sculptor" returns a choice between "Sculptor Group", "Sculptor Galaxy", "Sculptor Dwarf Elliptical", or "Sculptor Dwarf Irregular"; and "qso 1234" will suggest nomenclatures of the form "[HB89] HHMM+DDda" or "QS FN:NN" with explanations of their origin.

To underline the importance of the users' suggestions, every screen offers a menu option for users to leave comments on the system. Comments are reviewed and acted upon almost daily by the NED team. They include possible errors in the data, requests for new functions or data, and quality ratings of the service.

5.2 BATCH PROCESSING MODE

In this mode, a user e-mails a set of queries to nedbatch@ipac.caltech.edu, to be processed in a queue without human intervention. The results are written to file on a public access directory, and the user is notified by e-mail of the status of the processing, and the location and size of the results file. The queries have to be submitted following a specific format, a model of which will be mailed upon request by the primary interface.

5.3   SERVER MODE

This is the most recent interface to be offered by NED, and has therefore a more reduced access to NED data. It is designed to allow software developers to integrate a NED query into a computer application in a manner that requires no intervention from the user. For instance, a database of images accessible only by position could be enhanced by allowing users to enter names of galaxies, then have the software query NED directly for the position attached to the name, and use the resulting information to address the database; this provides a major enhancement at very low cost in development.

10–6   Architecture

The NED software resides on two platforms, a SUN Sparcstation 10 which supports all the user interface functions and the abstract database, and a SUN4 server which carries the search functions, and where most of the data are kept and managed. On the interface platform, the main software module which manages the session was written expressly for NED, but relies for screen input and output on a commercial package called JAM (for "JYACC Application Manager"), developed and marketed by JYACC. On the server side, all database management system functions are handled by SYBASE, a commercial software package that largely follows the relational database paradigm. Communication between platforms is handled by SYBASE client-server connections over the local IPAC network. This architecture offers many advantages in terms of performance, security, and manageability, and makes optimal use of the available platforms.

10–7   The Future

NED is still growing, both with new functions being added, and old ones becoming more sophisticated. In addition, the contents of NED continue to grow with more catalogs being folded into the Object Directory, more pointers to the literature being entered, more data — photometric and otherwise — being added, and more abstracts becoming available. In the long term, NED will certainly undergo much evolution, for it was designed with flexibility and ease of upgrades in mind. Enhancements presently envisaged call for more visualisation (plotting and text presentation) capabilities on the user interface, and an expanded suite of functions accessible via the server mode. Exploratory efforts are also under way in collaboration with J. Mazzarella to establish a Mosaic-based interface for NED to take advantage of the capabilities offered by the World-Wide Web system.

This is a time of transition, with the information revolution finally changing the way astronomers conduct their business, and electronic publishing becoming a reality. By their nature, database services are both symptoms and catalysts of that transition. NED is already playing its role in this arena, although it relies on the existing peer-review system for ensuring data quality, drawing its data primarily from the refereed literature. However, NED goes beyond these foundations by offering new modes of tapping into the literature; the growing collection of interconnected data, references and abstracts will make possible yet more innovative modes. By greatly simplifying certain queries, NED leads to new habits in research; projects that used to be dauntingly labor-intensive can now be undertaken by a lone researcher spending an afternoon at a terminal. Exploratory work that would have seemed "too risky" because of its considerable cost suddenly is worth the effort. How does NED contribute to astronomical research? By making it affordable to ask questions, NED paves a new high road to discovery.

Acknowledgments

Many people have contributed to NED, and continue to do so; we would like to thank Rick Ebert and Maneesh Sahani for their help in the software arena; we also recognize IPAC's support, especially through the computer systems group. We especially acknowledge the friendly collaboration of the SIMBAD project, Centre des Données astronomiques de Strasbourg. Last but not least, we thank all the authors and the National Space Science Data Center, who send us computer-readable versions of their data or catalogs.

NED is funded by the Science Operations Branch of the Astrophysics Division, Office of Space Science and Applications, NASA. This work is carried out by the Jet Propulsion Laboratory, California Institute of Technology, under a contract with the National Aeronautics and Space Administration.

References

[1] Abell, G.O., Corwin, H.G., Jr., and Olowin, R.P. (1989), ApJS, 70, 1. (ABELL)
[2] Abt, H.A. (1988), PASP, 100, 1567.
[3] Arakelian, M.A. (1975), Soobshch. Byurakan Obs. Akad. Nauk. Arm. SSR, 47, 1. (ARK)
[4] Arp, H.C., and Madore, B.F. 1987, "A Catalogue of Southern Peculiar Galaxies and Associations" (Cambridge: Cambridge University Press) (AM)
[5] Becker, R.H., White, R.L., and Edwards, A.L. (1991), ApJS 75, 1. ([BWE91])
[6] Burstein, D. and Heiles, C. (1978), ApJ, 225, 40.
[7] Caswell, J.L., and Crowther, J.H. (1969), MNRAS, 145, 181. (4C)

[8] Corwin, H. G. Jr., de Vaucouleurs, A., de Vaucouleurs, G. (1985), "Southern Galaxy Catalogue". (Austin, TX: University of Texas) (SGC)

[9] Dreyer, J.L.E. (1888), MemRAS 49, 1. (NGC)

[10] Gezari, D.Y., Schmitz, M., Pitts, P.S., and Mead, J.M. (1993), "Catalog of Infrared Observations", NASA RP-1294. (CIO)

[11] Goodwin, J.G., Metcalfe, N., and Peach, J.V. (1983), MNRAS, 202, 113. ([GMP83])

[12] Gower, J.F.R., Scott, P.F., and Wills, D. (1967), MmRAS, 71, 49. (4C)

[13] Gregory, P.C., and Condon, J.J. (1991), ApJS, 75, 1011. ([87GB])

[14] Helou, G., Soifer, B.T., and Rowan-Robinson, M. (1985), ApJL, 298, L7.

[15] Hewitt, A., Burbidge, G., (1989), ApJS, 69, 1 ([HB89])

[16] Hickson, P, Kindl, E., and Auman, J.R. (1989), ApJS, 70, 687. (HCG)

[17] Huchtmeier, W.K., and Richter, O.-G. (1989), "A General Catalogue of HI Observations of Galaxies" (New York: Springer-Verlag)

[18] IRAS Point Source Catalog, Version 2 (1988), Joint IRAS Science Working Group (Washington, DC: US GPO) (IRAS)

[19] Karachentsev, I. (1993), AN 314, 97. (FGC)

[20] Klemola, A., Hanson, R., and Jones, B. (1987), AJ 94, 501. (NPM1G)

[21] Lauberts, A. (1982), "The ESO/Uppsala Survey of the ESO(B) Atlas" (Garching bei Munchen: European Southern Observatory) (ESO)

[22] Lauberts, A. and Valentijn, E.A. (1989), "The Surface Photometry Catalogue of the ESO-Uppsala Galaxies" (Garching bei Munchen: European Southern Observatory) (ESO-LV)

[23] Metcalfe, N., Goodwin, J.G., and Peach, J.V. (1994), MNRAS, 267, 431. ([MGP94])

[24] Moshir *et al.* (1992), "Explanatory Supplement to the IRAS Faint Source Survey, Version 2", JPL D-10015 (Pasadena: JPL) (IRAS F)

[25] Nilson, P. (1973), "Uppsala General Catalogue of Galaxies", Acta Universitatis Upsaliensis, Ser. V:A vol. 1 (Uppsala: Reg. Soc. Scient. Ups.) (UGC)

[26] Nilson, P. (1974), "Catalogue of Selected non-UGC Galaxies", Uppsala Astronomical Observatory Report No.5. (UGCA)

[27] Palumbo, G.G.C., Tanzella-Nitti, G., and Vettolani, G. (1983), "Catalogue of Radial Velocities of Galaxies" (New York: Gordon & Breach)

[28] Parkes Catalogue (1990), the Southern Radio Source Database, ver. 1.01, compiled by A.E. Wright and R.E. Otrupcek. Magnetic tape and preprint, Australia Telescope National Facility. (PKS)

[29] Pilkington, J.D.H., and Scott, P.F. (1965), MmRAS, 69, 183. (4C)

[30] Saito, M. *et al.* (1991), PASJ, 43, 449. (CGMW)

[31] Sandage, A. (1961), "The Hubble Atlas of Galaxies" (Washington, DC: Carnegie Institution of Washington)

[32] Schmidt, M., and Green, R.F. (1983), ApJ, 269, 352. (PG)

[33] Schombert, J.M., and Bothun, G.D. (1988), AJ, 95, 1389. ([SB88])

[34] Soifer, B.T. *et al.* (1984), ApJL, 283, L1.

[35] Takase, B., and Miyauchi-Isobe, N. (1993), Publ. Natl. Obs. Japan, 3, 169. (KUG)

[36] de Vaucouleurs, G., de Vaucouleurs, A., and Corwin, H.G., Jr. (1976), "Second

Reference Catalogue of Bright Galaxies" (Austin: University of Texas Press) (RC2)

[37] Vorontsov-Velyaminov, B. et al. (1962-74), "Morphological Catalogue of Galaxies", in 5 parts (Moscow: Moscow State University) (MCG)

[38] White, R.L., and Becker, R.H. (1992), ApJS, 79, 331. ([WB92])

[39] Zwicky, F. et al. (1961-68), "Catalogue of Galaxies and of Clusters of Galaxies", in 6 volumes (Pasadena, CA: California Institute of Technology) (CGCG)

Access Pointers

[→1] NED WWW page:
    http://www.ipac.caltech.edu/ipac/projects/ned.html

[→2] NED telnet access: telnet://ned@ned.ipac.caltech.edu/

11

LEDA: The Lyon-Meudon Extragalactic Database

G. Paturel, I. Vauglin, H. Andernach, R. Garnier,
M. C. Marthinet, C. Petit, H. Di Nella, L. Bottinelli,
L. Gouguenheim & N. Durand

11–1   Origin of LEDA

In 1981 our scientific team was involved in the management of large amount of HI-data. The goal, at this time, was only our own scientific research. In a collaboration with G. and A. de Vaucouleurs we started to add optical information to our first HI-catalog. It was our first contact with the difficult problem of cross-identifying two catalogs and merging them. To face this problem, one of us (GP) decided to rationalize the management by creating a database. It was created in 1983 at Lyon Observatory. LEDA is thus the oldest Extragalactic Database. Since this time, the database has been continuously updated and improved.

11–2   LEDA's data

The main idea is to collect raw measurements (directly from published observations) and to archive them. From these raw measurements, mean homogenized data is calculated in the same spirit as de Vaucouleurs did when he published his famous series of Bright Galaxy Catalogues (RC1, RC2, RC3). In fact, RC3 was created using LEDA database. Additional model-dependent parameters are calculated too.

The most important observed parameters which have been published in the literature for galaxies are collected and archived in LEDA: cross-identifications, coordinates, morphological description, diameters, axis ratio, surface brightness, flux at different wavelength, 21-cm line width, central velocity dispersion, group membership, radial velocities. Each parameter is

D. Egret and M. A. Albrecht (eds.), Information & On-Line Data in Astronomy, 115–126.

given with bibliographic reference and basic characteristics of the measurement. The full list of available parameters is given in the annex below with a short description of the method used for their calculation.

```
                LYON-MEUDON EXTRAGALACTIC DATABASE
                = MEAN  HOMOGENEIZED  PARAMETERS =
                =================================
==============================================================================
PGC 002557= NGC    224       = UGC    454      = MCG  7- 2- 16  = MESS   31
           = CGCG 535- 17    =
------------------------------------------------------------------------------
    al     de            12        b2      sgl         sgb        pa      lgg group
B004000.1+405943*      121.17   -21.57   336.45      12.55        35.0    LGG    11
J004244.4+411608       (J=2000)
------------------------------------------------------------------------------
typ        t      lc   logd25 logr25   bt     m21    mfir    bvt      ubt
Sb        3.0    2.0   3.28    .48    4.17   6.15    5.66    .92      .50
           .4     .6    .01    .02     .04    .08
------------------------------------------------------------------------------
   w20    w50    vrad   vopt  brief   bri25
   536    510    -301   -296  22.10   22.98
     7      7       4     13    .44
------------------------------------------------------------------------------
  incl     ag     ai   logdc   btc    bvc    ubc   bve    ube
  77.4    .33    .59   3.19    3.25    .76    .35
------------------------------------------------------------------------------
  logs   logvm  lambda  vgsr    vlg    vvir   v3K   mucin  mupar   mabs
  2.223  2.397    .44   -123    -14    -104   -576          24.96 -21.71
------------------------------------------------------------------------------
Q = quit    H = help    S = save   <RETURN> = continue  :
```

Table 11-1: Mean homogeneized parameters for NGC 224 from a LEDA session.

11-3   General description of LEDA

To get the connection via internet enter:

```
telnet lmc.univ-lyon1.fr
login :leda
```

Two interfaces are supported: VT100 for a simple connection in ASCII mode and an X11 interface with full graphical capabilities. An *Instructions-for-use* file is accessible on-line.

The structure of the main menu is the following:

```
MAIN MENU:  Single object      : Explore raw data
                                : Explore mean data
            Several objects     : Select from a list of names
                                : SQL-like selection
            2D-information      : Charts
```

```
                              : Images
        Information           : Instructions for use
                              : News
                              : LEDA's team
                              : Status
        Quit                  : Give comments
```

Except for the SQL-like option, *no prior knowledge is necessary to use LEDA.* Generally, you just have to follow on-line instructions. Even the designation of an object is interpreted by a powerful artificial intelligent software providing a large flexibility to users. For instance, the designation of an object can be given as a literal name (*e.g.* LMC) or as a common designation (*e.g.* NGC224 or n224 ...) or as equatorial coordinates in a free format (as far as it is understandable by user).

The result of each request is sent back via e-mail provided that your e-mail address is correct.

Now let us give more details about some powerful options.

11–4 SQL-like language

This option is the most powerful tool provided by LEDA. The Structured Query Language (SQL) has been developed by IBM to formulate even very complex queries. The LEDA query language is somewhat similar. The principle consists in giving a sentence which describes the request. The structure of this sentence is always the same:

```
select  (parameters desired for output)  where  (conditions)  end
```

The parameters you desire in the output file are simply given as a list of parameter names (or a combination of them built with operands given hereafter), each name being separated by a comma.

```
select  ident1,al2000,de2000,logd25  where  (conditions)  end
```

The conditions are built using operators like >, <, = and operands (+, -, *, /, ^ (exponent), & (exist), not() (negation), log(), ln(), exp(), sqrt(), sin() cos(), tan(), abs()). Note that you must put character strings between apostrophes (') (*e.g.* typ='E'). Here are some examples:

```
select ident1, logd25
    where (w20>200 and w20<300) end
select ident1
    where &w20 or &w50 end
select pgc, al2000, de2000
```

```
    where (bt+5*logd25)<20 end
select pgc, al2000, de2000
    where al2000<3.0 and al2000>2.0 end
select pgc, typ
    where typ='E' and bt<10 end
select pgc, typ, logs
    where typ='E' and not(&logs) and bt<12 end
select pgc, (bt+5*logd25)
    where (bt+5*logd25)<20 end
select pgc
    where ring='R' and multiple='M' end
select pgc, abs(w20-w50)
    where &w20 and &w50 and abs(w20-w50)>100 end
```

The most important thing to know is the name and definition of each parameter. The authorized parameter names are given in the annex below.

11–5   Batch mode

If you want to make a study (*e.g.* extract data, make charts, plot on a Flamsteed projection) from a list of galaxy names or galaxy positions, it is not necessary to keypunch these names or positions in an on-line connection. Just send your list of galaxy identifiers (names or 1950-coordinates) to

`ledamail@lmc.univ-lyon1.fr`

and give a *Subject* depending on the task to be accomplished. You will receive in reply to your mail a file with the result. Several tasks are available. Let us describe what you will receive in reply to each of them.

5.1   Subject: LIST OR LISTALL

You will receive an ASCII file with the main parameters needed for good identification of the galaxies of your list. These parameters are: coordinates, alternative names, apparent diameter and axis ratio, apparent magnitude, radial velocity. With the subject LISTALL you will receive all the parameters (see the on-line documentation for their formats and units).

5.2   Subject: FLAMEQ (OR FLAMGA OR FLAMSG)

You will receive a postscript file with a Flamsteed equal area projection of your galaxies in equatorial coordinates (FLAMEQ), galactic coordinates (FLAMGA) or supergalactic coordinates (FLAMSG), depending of the Subject you selected. To print this file just send it to a laser printer working with postscript language.

5.3 Subject: CHART

You will receive a postscript file with charts (maximum of 20 charts at a time), and a corresponding ASCII file with data. Note that two lines must appear in the input file for each object: **first line:** identifier (Name or 1950-RA.DEC) **second line:** radius in arcmin (0 for PSS or ESO scale): For instance, the following file:

```
m31
0
12 19 21.4 4 44 58
100
```

will create two charts (in one file): the first chart will be centered on Messier 31 with the scale of the Palomar Sky Survey (PSS). The second one will be centered at the position RA1950=12h19m21.4s DEC1950=04d44'58" with a radius of 100'.

11–6   2D–Information

We will describe how 2D–information can be retrieved on-line via an X11-interface. Certainly, this information can also be obtained as postscript files from a VT100 interface or batch mode.

A chart at a given position and for a given radius can be drawn directly on screen with an X11 terminal (see figure 11–1). The list of galaxies appears in a scrolling window. Stars are superimposed; they can be removed if necessary. It is possible to zoom at any position. An ESO/PSS automatic scaling is provided allowing the user to produce transparencies for easy identification on ESO or PSS charts. The save push-button option sends a postscript file for printing to a laser printer. This postscript chart is provided with an ASCII file, giving the main parameters of galaxies of the field.

Many images of galaxies (more than 34,780 in April, 1994) can also be displayed on the screen (see figure 11–2). All galaxies appearing in the field are listed in a scrolling window, in such a way that, clicking on a galaxy name, will put the arrow on the corresponding galaxy. Like for charts, these images can be saved and printed on the user's laser printer.

11–7   Future developments

Due to the success of the batch mode (ledamail) we are developing new options. The SQL-like language will be available soon also via ledamail. The

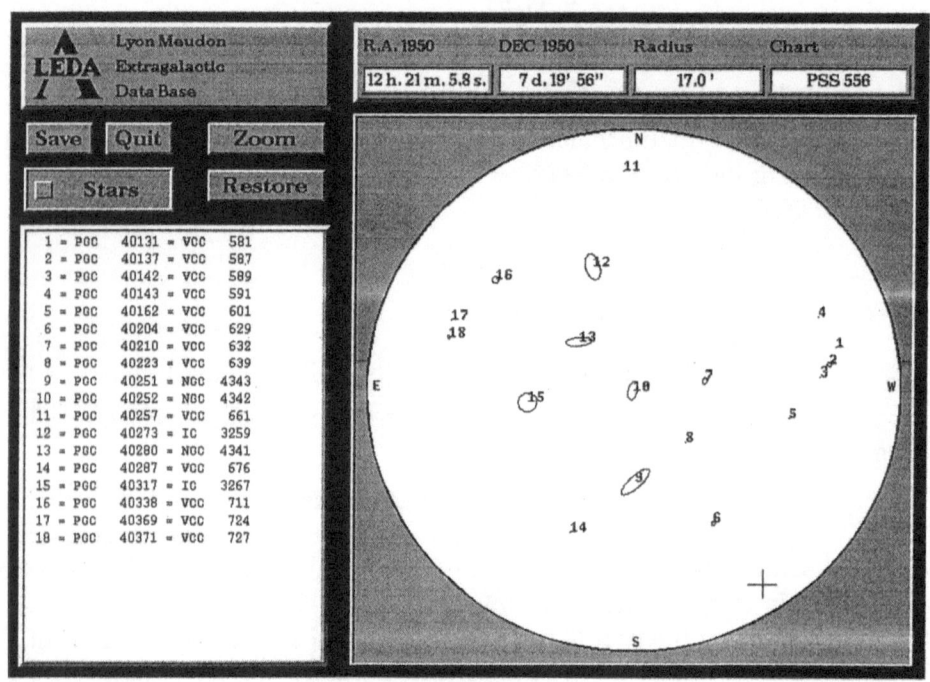

Figure 11–1: The LEDA chart display.

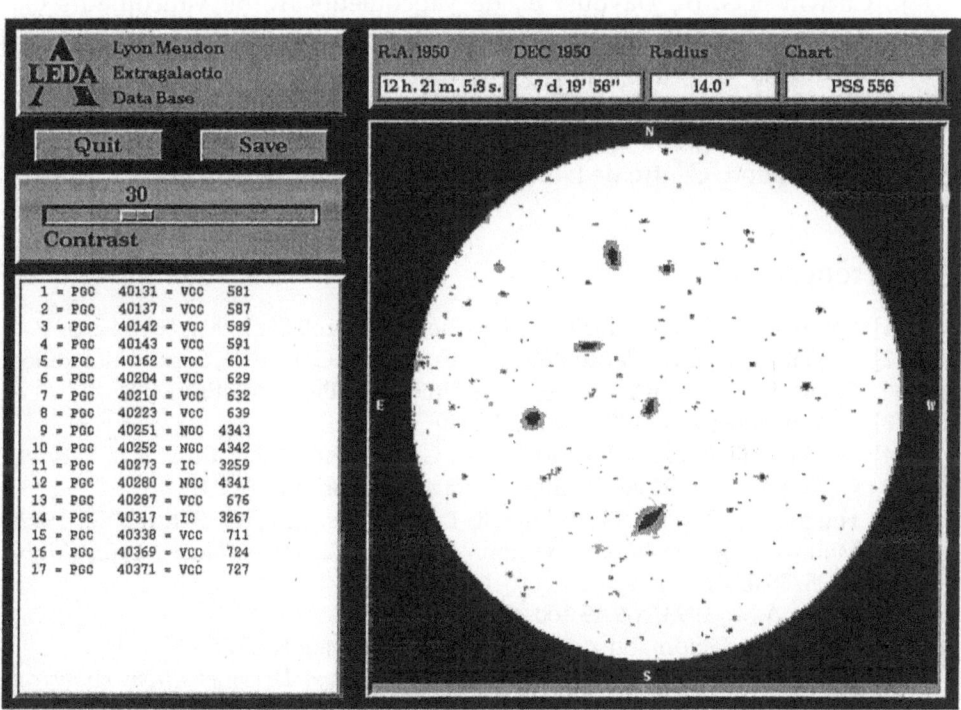

Figure 11–2: The LEDA image display.

involvement of LEDA in the near infrared DENIS Survey (see related chapter in this volume, page 115) will provide users with more data and images. The database is regularly updated and the most interesting data is distributed on CD-ROM for local use with small personal computers.

Acknowledgements

LEDA has been developed with the help of many people: Becker M., Buta R.J., Corwin H.G. Jr., Davoust E., de Vaucouleurs A., de Vaucouleurs G., Durand S., Fouqué P., Hallet N., Kogoshvili N., Legrand C., Mamon G., Miyauchi-Isobe N., Odewahn S., Paturel Ph., Prugniel Ph., Simien F., Takase B.. LEDA has been supported by the following Institutions: INSU, CNRS, DRED, Conseil Regional Rhône-Alpes, Observatoire de Lyon, Observatoire de Paris-Meudon, Centre de Données astronomiques de Strasbourg.

References

[1] Bottinelli, L., Gouguenheim,L., Fouqué, P., Paturel, G., 1990 A&AS 82, 391

[2] de Vaucouleurs G., Vaucouleurs A., Corwin, H.G. Jr, 1976, Second Reference Catalogue of Bright Galaxies, Texas University Press, Austin (RC2)

[3] de Vaucouleurs G., 1979, ApJ. 227, 380

[4] de Vaucouleurs, A., de Vaucouleurs, G., Corwin, H.G.Jr, Buta, R.J., Paturel, G., Fouqué, P. 1991 ed. Springer-Verlag New-York Inc. (RC3)

[5] Fouqué, P., Durand, N., Bottinelli, L. ,Gouguenheim, L., Paturel, G., 1992, Monographies de la base de donnees extragalactiques, No.3 , Lyon (ISBN 2.908288.05.2)

[6] Garcia A.M., 1993, A&AS 100, 47

[7] Heidmann, Heidmann, de Vaucouleurs, 1971, MemRAS 75, 85

[8] Lubin P., Villela T, 1986, in *Galaxy Distances and Deviations from Universal Expansion*, ed. B. Madore and R. Tully, Dordrecht: Reidel, p169

[9] Paturel,G., Fouqué,P., Buta, R.J., Garcia, A.M., 1991, A&A 243, 319

[10] Paturel G., Fouqué P., Bottinelli L., Gouguenheim L., 1989a, A&AS 80,209

[11] Paturel G., Fouqué P., Bottinelli L., Gouguenheim L., 1989b, Principal Galaxy Catalog, Monographies de la base de donnees extragalactiques, No.1 , Lyon (ISBN 2.908288.00.1) (PGC)

[12] Paturel G., Bottinelli L., Gouguenheim L., 1994, A&AS (in press)

[13] Tully, R., Fouqué, P., 1985, ApJS 58,67

[14] Yahil, A, Tammann, G.,A., Sandage, A., 1977, ApJ 217, 903

Access Pointers

[→1] LEDA: http://www-obs.univ-lyon1.fr/base/home_base.html

[→2] LEDA Session: `rlogin://leda@lmc.univ-lyon1.fr/`

Annex: Glossary of LEDA'S astrophysical parameters

(Please, check the on-line documentation, because definitions and methods of calculation may change without warning)

*al*1950, *de*1950 : 1950-Right ascension and declination in decimal hours and degrees, respectively

ipad : flag to tell that *al*, *de* have an accuracy better than 10"

*al*2000, *de*2000 : 2000-right ascension and declination in decimal hours and degrees, respectively

*l*2, *b*2 : galactic longitude and latitude in degrees

sgl, *sgb* : supergalactic longitude and latitude in degrees (pole and origin conforms to the RC2 reference system)

pgc : LEDA identifier (PGC number when it exists)

*ident*1 : Alternative name given according to a hierarchy: NGC; IC; UGC; ESO; MCG; UGCA; MARK; DDO; CGCG; FAIR; ARAK; UM; ZW; KUG; HICK; VCC; IRAS; WEIN; VIII; ISZ; POX; KAZA; KARA; CR; RB; DRCG; SAIT; KCPG; FGC; FGCE; FGCA For the description of names see Paturel *et al.* 1989a, 1989b

pa : position angle (in degrees measured from North towards East)

typ : mean morphological type

ring : 'R' for Ring galaxies, otherwise ' ' (blank)

multiple : 'M' for multiple galaxies, otherwise ' ' (blank)

compact : 'C' for compact, 'D' for diffuse, otherwise ' ' (blank)

t : mean morphological type code according to RC3 (de Vaucouleurs et al. 1991) $(E = -5 ; SO = -2 ; Sa = 1 ; Sb = 3 ; Sc = 6 ; Sm = 9 ; Irr = 10)$

st : mean error on type code *t*

lc : luminosity class coded according to RC3 (I=1 ; I-II=2 ; II=3 ...) and corrected for inclination effect.

slc : mean error on luminosity class *lc*

*logd*25 : log of the apparent major axis of the galaxy at the $25mag.arcsec^{-2}$ surface brightness level. *d*25 is in 0.1'.

*slogd*25 : mean error on *logd*25

*logr*25 : axis ratio $logr25 = log(d25/b25)$ where *d*25 and *b*25 are major and minor axes, respectively

*slogr*25 : mean error on *logr*25. The last four quantities are calculated according to Paturel et al. 1991.

brief : mean effective surface brightness in B. The definition is the one given in the RC3 catalog but it is expressed in $mag.arcsec^{-2}$. $(bt + 0.753) + 5logde - 5.26 + 8.891$, where bt is the total B magnitude, and de the effective diameter in 0.1′.

sbrief : mean error on *brief*

bt : total (or asymptotic) blue magnitude

sbt : mean error on *bt* The last two quantities are derived from Paturel *et al.* 1994 (in press).

ubt : total (U-B) color from aperture photoelectric photometry

bvt : total (B-V) color from aperture photoelectric photometry

ube : mean effective (U-B) color from aperture photoelectric photometry

bve : mean effective (B-V) color from aperture photoelectric photometry The definition of the last four quantities is taken according to the RC3

w20 : mean 21-cm line width at 20% of the maximum (in km/s)

sw20 : mean error on *w20* (in km/s)

w50 : mean 21-cm line width at 50% of the maximum (in km/s)

sw50 : mean error on *w50* (in km/s) The last four quantities are calculated according to Bottinelli *et al.* , 1990. They are corrected only for instrumental resolution effect.

logs : decimal log of the central velocity dispersion (in km/s)

slogs : mean error on *logs*

m21 : magnitude corresponding to the 21-cm line area. The definition of $m21$ conforms to the RC3 definition: $m21 = 16.6 - 2.5logSH$ where SH is expressed in $10^{-22}W.m^{-2}$; $m21$ is corrected for beam filling effect and is calculated according to Bottinelli *et al.* , 1990.

sm21 : mean error on *m21*

mfir : far-infrared magnitude derived from the IRAS Point Source Catalog. The definition is the following: $amfir = -20 - 2.5log(FIR)$ where $FIR = 1.26[2.58f(60) + f(100)] \times 10^{-14}$ being $f(60)$ and $f(100)$ the IRAS fluxes at 60μ and 100μ respectively. FIR is expressed in $W.m^{-2}$.

vrad : mean heliocentric radial velocity from 21-cm line measurements in km/s.

svrad : mean error on *vrad* The last two quantities are derived according to Bottinelli *et al.* , 1990

vopt : heliocentric mean radial velocity from optical measurement.

svopt : mean error on *vopt* The last two quantitites are derived according to Fouqué *et al.* , 1992

v : mean heliocentric velocity in km/s (weighted mean of vrad and vopt). The weights are calculated as the inverse of the square of the mean error.

sv : mean error on *vm*.

lgg : Lyon's galaxy group number (LGG) according to Garcia, (1993).

ag : galactic absorption in blue magnitude. This absorption is calculated according to the RC2.

ai : total internal absorption in blue magnitude It is calculated according to Tully and Fouqué 1985

incl : inclination between the line of sight and the polar axis of the galaxy. It is calculated from axis ratio through the ellipsoidal model: $sini = \sqrt{(1-q^2)/(1-q_o^2)}$ where $q = 1/r25$, q_o being a function of morphological type code $q_o = .43 + 0.053t$ and $q_o = 0.38$ for $t > 7$ (Fouqué, private communication)

a21 : HI self-absorption according to Heidmann *et al.* , (1971) with coef. $\tau = 0.031$

lambda : luminosity index corrected for inclination effect. This index has been defined by G. de Vaucouleurs 1979 as: $(t+lc)/10$ where t is the morphological type code and lc the luminosity class corrected for inclination effect.

logdc : decimal log of the corrected apparent diameter dc, where dc is in 0.1′. *logdc* is corrected for galactic absorption and inclination effect using the following relation: $logdc = logd25 - 0.235logr25 + ag(0.081 - 0.016t)$

btc : corrected total apparent blue magnitude B_o corrected for galactic absorption, internal absorption and redshift effect The corrected magnitude is calculated using the relation: $btc = bt - ag - ai - K.v$ where v is the heliocentric radial velocity and K a function of morphological type code t (according to RC2)

ubtc : (U-B)T color, corrected according to RC2

bvtc : (B-V)T color, corrected according to RC2

bri25 : mean surface brightness at the $25mag.arcsec^{-2}$ surface brightness level. This mean surface brightness is expressed in $mag.arcsec^{-2}$.

logvm : log of the maximum velocity rotation vm from 21-cm line width $w20$ and $w50$; vm is corrected for inclination effect and for internal dispersion according to Tully and Fouqué (1985).

m21c : corrected HI magnitude $m21c$. A correction of self-absorption is made according to Heidmann *et al.* (1971).

hic : hydrogen index HI defined following: $hic = m21c - btc$

vlg : mean radial velocity corrected for solar motion towards the centroid of the Local Group. The correction is made according to Yahil et al. 1977

vgsr : radial velocity corrected for solar motion toward the galactic standard of rest. The correction is made according to RC3. (232.3 km/s toward $l2 = 87.8d$; $b2 = 1.7d$)

vvir : radial velocity corrected for infall of the Local Group toward the Virgo cluster. The adopted infall velocity is 150 km/s toward supergalactic position $sgl = 104degrees$, $sgb = -2degrees$ (Virgo center)

v3k : radial velocity corrected to the reference frame of the 3K radiation according to Lubin and Villela 1986. (solar motion 360 km/s toward $al1950 = 11.25h$; $de1950 = -5.6deg$) The last four parameters are expressed in km/s *mucin* : distance modulus calculated from kinematical velocity vvir using a Hubble constant of $75 km.s^{-1} Mpc^{-1}$. The distance modulus is not calculated for *vvir* < $500 km/s$.

mupar : mean distance modulus calculated from parameters using a weighted mean of all distance criteria (logVm, Lambda-c, dispv, U-B). The adopted relation are the following:

$mupar = btc + 6.5.logvm + 6.3 + 0.02 * v/1000$

$mupar = btc - 3.4.lambda + 22.9 + 0.16 * v/1000$

$mupar = btc + 6.2.logs + 5.9 + 0.16 * v/1000$

$mupar = btc + 4.4.ubtc + 17.8 + 0.26 * v/1000$

The velocity correction aims at correcting for Malmquist bias. Obviously such a correction is only a tentative improvement.

mabs : absolute magnitude derived from the adopted distance modulus *mucin* and corrected apparent magnitude btc.

12

The Database for Galactic Open Clusters (BDA)

Jean-Claude Mermilliod

12-1 Introduction

The database for stars in galactic open clusters (BDA) has been developed since 1987 at the Institute for Astronomy (University of Lausanne). The extensive collection of observational data covers most significant domains and concerns about 100,000 stars in some 500 NGC, IC and anonymous clusters.

It includes measurements in most photometric systems in which cluster stars have been observed, spectroscopic observations, astrometric data, various kinds of useful information, and extensive bibliography. Maps for about 200 clusters have been scanned and included in the database. The greatest effort has been spent in solving the identification problems raised by the definition of so many different numbering systems.

The database not only aims at storing data but also at offering a versatile working environment covering many aspects of the study of open clusters. It provides tools to compare the data and plot photometric diagrams. The facilities include the on-line computation of isochrones. It is most suitable for use on a workstation. Independently of the application software proposed, the main utility of the database lies in the extensive data collection brought into uniform numbering systems. The BDA presents a clear report of the present status of observations.

12-2 Context

About 1200 galactic open clusters are known and half of them have been observed so far, in at least one photometric system. The number of stars per cluster goes from several tens for the poorest objects, to several thousands for the most prominent clusters. Modern observations of open clusters developed

D. Egret and M. A. Albrecht (eds.), Information & On-Line Data in Astronomy, 127–138.

very rapidly after the definition of the UBV photoelectric system (Johnson & Morgan 1953). These observations produced a number of colour-magnitude (Hertzsprung-Russell) diagrams which made fundamental contributions to the understanding of stellar evolution. At the same time photographic photometry allowed to observe larger areas and reach fainter stars. Additional information, mostly spectroscopic, was gradually obtained, first for stars in the nearby clusters and later in more distant clusters, thanks to the existence of larger telescopes and more efficient detectors. Two-dimensional detectors are best adapted for the observations of star clusters and CCD observing is today becoming the preferred technique. It replaces both the photoelectric and photographic photometry.

Data compilations started already in 1972 at the Institute for Astronomy (University of Lausanne, Switzerland) with the financial support of the Swiss National Funds for Scientific Research (FNRS). Mermilliod (1976) published the first catalogue of UBV photometry and MK spectral types. The third version was announced ten years later (Mermilliod 1986). The systematic determination of cross-references between the many numbering systems in a cluster was the basic work which made the realisation of the data collections possible. Several catalogues were distributed by the Strasbourg Data Center (CDS). The files remained on magnetic tapes until the installation of Unix workstations and large disks in our institute made it possible to keep the data on-line. These compilations were discontinued in their older form and the data were organised in a database designed in March 1987 (Mermilliod 1988a, 1988b).

The database has been developed not only to be an efficient tool to store and retrieve data, but also to provide a versatile environment to analyse the data and study open clusters considered as interesting astrophysical objects worth a systematic study.

12–3   The astronomical contents

The database tries to collect all published data for stars in open clusters that may be useful either to determine the star membership, or to study the stellar content and properties of the cluster. The data are usually recorded in their original form, with an indication of the source, but also as averaged values or selected data when relevant. The mean values for UBV (photoelectric, photographic or CCD) are not kept in the database, but can readily be computed. It presently contains:

> ▷ **astrometric** data: coordinates, rectangular positions, and some proper motions;

▷ **photometric** data in most systems in which cluster stars have been observed;

▷ **spectroscopic** data: spectral classification, radial and rotational velocities;

▷ **bibliographic** information;

▷ **specific** information: (a) membership probabilities, (b) orbital elements of spectroscopic binaries, (c) remarks on peculiarity, variability, duplicity, (d) identifications of double star components, (e) cross-identifications with astronomical catalogues, (f) list of red giants in the cluster field and of non-member stars.

Table 12–1 summarises the May 1994 contents of the database for the main data types. The columns give successively the data type, the number of clusters, measurements and stars concerned. A progress report on the introduction of new data has been published by Mermilliod (1992a).

Apart from the published data found in the literature, the database contains also information that was accumulated in the past or prepared especially for it to enhance its capabilities:

▷ Cross-reference tables contain the basic cross-identifications between the various numbering systems in a cluster. A large part of the work has been done by the author.

▷ Rectangular (x, y) positions (usually in arbitrary units) have been collected from the literature or measured on published photographs with a digitizing tablet. A file collects the information relevant to the scale of the (x, y) positions and the sources of the data.

▷ Published photographs defining cluster numbering systems have been scanned and included in the database. About 200 maps have already been processed.

The access to general or bibliographic information is considered as being a very important facility that the database should offer. The catalogues already available in computer-readable form have been installed in the database and an additional bibliographic service covering the recent literature has been developed in another style to make the retrieval easier and more efficient. The various facets are provided by:

▷ the compilation of cluster parameters by Lyngå (1987), which provides the global information available on open clusters;

▷ the bibliography compiled by Alter et al. (1970) and its first supplement (Ruprecht et al. 1981), which collect the references from the remote past up to 1973;

Subjects	Number of		
	Clusters	Meas.	Stars
Identifications	365		9876
Transit Table	188		68498
Coordinates	441	30621	27879
Positions	441		39424
Positions (x,y)	384		112607
Double stars	193	1662	1230
UBV photoelectric	412	30309	20600
UBV photographic	267	91338	71779
UBV CCD	69	30409	28322
RGU (pg)	74	10191	10191
Geneva 7-colors	184		4306
uvby measures	155	5329	3780
uvby mean	155		3423
uvby Eggen	42	876	775
Hβ measures	216	5567	3707
Hβ mean	216		3312
DDO	126	950	769
Washington	60	533	507
Walraven	58	1432	1382
RI (Eggen)	14	118	118
RI (Cousins)	23	869	847
RI (Cousins) CCD	8	3149	3112
Vilnius	31	747	698
MK types	271	7746	4637
MK types (selected)	162		3974
HD types	299		8584
Vsini	79	2222	1592
RV mean	62	1791	1604
RV individual	183	29542	3442
RV GPO	10	568	568
Orbits	35	199	175
Probability (μ)	59	21574	21574
Remarks	234	3703	3146
gK stars	231		3291
Bibliography (Alter)	1200		
Bibliography (69-94)			3000

Table 12–1: May 1994 database content

▷ the recent bibliography, covering the years 1969 to the present day, which has been developed for the database. This bibliography is based on chapter 153 of the Astronomy and Astrophysics Abstracts and the recent references are regularly entered in the computer. The bibliography search is based on keywords. The most obvious one is simply the cluster name, but many others can be used. Abstracts are not yet included in the database;

▷ the information on ongoing work, generally extracted from the reports of observatories and AAS abstracts is also available from the database.

With the exception of the photoeletric photometric data in the UBV, uvby and Geneva systems, most data contained in BDA cannot be found in Simbad. Thus the present database is not a subsample of Simbad data set.

12–4   Organisation of the data

The database is not a traditional relational database management system, but rather an advanced file management system. Clusters, but not stars, form the basic unit of the database and the structure has been designed to provide a natural working environment. The whole data set for a cluster forms a special relational set, because one key —the star designation— is common to all files.

The database structure uses the directory hierarchy supported by the Unix system. The main directory is the database itself. It contains several sub-directories: description of the database, help information, references, bibliography, programs, shell scripts. The clusters are collected in parent directories according to the source catalogues (NGC, IC or anon). Each cluster defines an independent directory identified by its name and containing the available data in distinct files, one for each data type. This structure allows easy inclusion of any new data type.

Whenever possible, the records of the various data files have the same structure: star identification, source, data. The files are organised sequentially and compressed and, within the files, the entries are sorted by star number and source reference. Due to the small size of many files, there is no need for indexing or direct access. The star identification is the main key to access the data, but it is also possible to use filters based on the bibliographic references or astrophysical parameters. The files are uncompressed on the fly and the data sent through a pipe.

The present database structure is thought of as a first step in the organisation of cluster data and bibliography. When enough data analysis has been performed and only one set of data for each type will be available for each star, it may perhaps be more convenient to adopt another structure and collect all the data in one file for each cluster.

12–5   Command style and software

The file organisation and their small size (a few hundred records, often less) make their use very rapid with Unix tools. The commands executing data requests are often Bourne shell scripts. They present a Unix style, with many options which have the same meaning for most commands and use the file corresponding to the requested data type in the current cluster directory. As a first policy, the command names are identical to the data type they handle: the command ubv will deal with the UBV data and mk, with the MK spectral types. The command syntax is rather simple and examples are given by Mermilliod (1988a). Due to the development of numerous options, the answer time became longer and an opposite philosophy has been developed: the command name is related to the action to be performed. The arguments are the data type and star numbers, like in Measure ubv 1 7 9, where 1 7 9 are the star numbers for which UBV data are requested.

Fortran codes are used for scientific applications because most of them were developed before the birth of the database, and C codes have been written for system programming (pull-down menus, graphics display). A graphical user interface under Sun's Xview is being developed providing a choice panel for the data types and command or menu buttons.   The application software use routines from the Numerical Recipes library (Press et al. 1986). The graphics have been built on the SM (2.3) graphics software package written by Robert Lupton and Patricia Monger and is based both on SM macros and the SM C function library. This software is not distributed with the database and should be installed separately on the host computer.

12–6   Facilities offered by the database

The database offers several working facilities and some are described below. Three examples are discussed by Mermilliod (1992a).

> ▷ **Data analysis:** Data analysis is an important step before starting any study. Tools have been developed to compare data coming from different sources. This concerns essentially the UBV system because the UBV data represents a large fraction of the published photometric data. The other photometric systems seldom present two or more different sources of data.

> ▷ **Colour-magnitude diagram manipulation:** The main tool of the database allows to plot the various diagrams that can be built in the UBV

and Geneva photometric systems. It offers a large spectrum of facilities, like fitting sequences or computing isochrones. Extensions have to be developed for other photometric systems (uvby, Walraven and others). Figure 12–1 shows an example of the colour-magnitude diagramme obtained for the young open cluster NGC 2287 from the UBV data contained in the database.

▷ **Database contents:** The numbers of stars observed and of measurements available are kept in separate files. This information may be used to prepare cluster samples by selecting the clusters that have enough data to perform a study, or very few data, to prepare an observing program.

▷ **Cluster catalogue:** A file, based on Lyngå's (1987) list of open clusters, contains the cluster identifications, coordinates and parameters (distance, reddening, age, earliest spectral type and diameter). It is possible to make selections on any parameter in this table to prepare cluster samples.

▷ **Technical data:** The technical information on instruments and data acquisition systems may also prove important for the characterisation of the data contained in the database. Pioneering work has been done by van Leeuwen (1985) who collected the information on telescopes and plates related to the proper motion studies in open clusters. This information is available in the database. The collection of similar information on telescopes and spectrographs used for radial velocity determination has been started. It would also be necessary to collect the same information for CCD photometry.

▷ **Miscellanous facilities:** To avoid remembering many commands and their options, a number of menus have been written which group most actions concerning a given subject. Those built up so far are generally related to the maintenance and development of the database, and were especially designed to facilitate the introduction of new data. In this category, one finds:

1. the management of *coordinates:* it was made to determine coordinates of stars in open clusters starting from (x,y) positions in arbitrary units and include them in the database;

2. the management of *rectangular positions:* (x,y) positions are now often published with CCD data. In addition many published charts have been measured and the results included in the database;

3. the management of *cross-references:* to maintain the principles of the database, it is necessary to determine the cross-references between any new numbering system and those already existing.

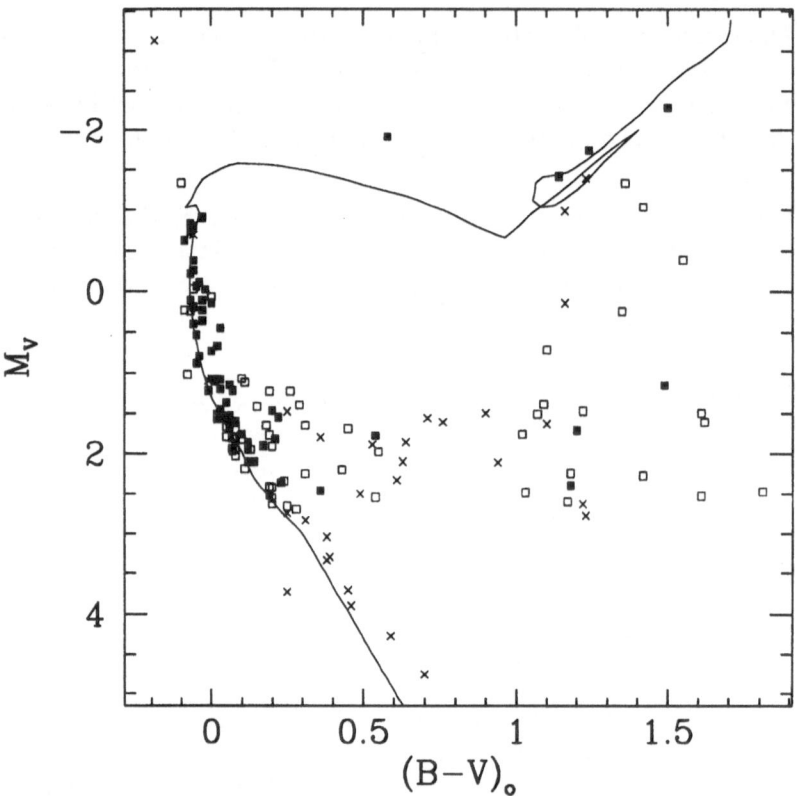

Figure 12–1: Colour-magnitude diagram for the open cluster NGC 2287. The observations are corrected for a distance-modulus m-M = 9.20, which takes into account
the metallicity (Fe/H = 0.09) given by Lyngå (1987), and a reddening E(B-V) = 0.01.
The isochrone (continuous curve) is based on Schaller et al. (1992) models for logt =
8.37. The various symbols indicate membership according to the published proper
motions (Ianna et al. 1987). Members (at 50% level) are designated by filled squares,
non-members, by open squares and stars without information, by crosses.

12–7   Access mode

Due to the specific structure of the database, it is presently more convenient to have a copy of it on a workstation. The database is maintained on a Sun Sparc workstation and its total size is about 35 MB. The dissemination of various copies may raise the problem of the simultaneous existence of several revisions. However, this problem is solved by the transfer of the modified or new data files grouped in a tar-file over the Internet. Such a system is already used to maintain a copy of the database at the Meudon observatory in Paris on a DEC workstation, which offers additional possibilities for distribution to other machines. Access through the World-Wide Web is being examined. Hypertext description is being written and some database consultation will be provided. There is also a project to access the database through the ADS facility.

12–8   User's profile

Even if the proposed software is not used, the utility of the database lies in the extensive collections of data brought into a uniform numbering system. The database can therefore be useful not only to astronomers working in the field of open star clusters, but also to students for a variety of work, because of the easy access to the data and facility for implementing users' programs.

Copies of the database have been sent to several colleagues in various countries. Other colleagues and several students working on their Ph.D. have also asked for specific data or bibliographic information by E-mail.

12–9   Scientific use

The realisation of a new atlas of colour-magnitude diagrams based on improved and homogeneous cluster parameters was the first scientific use of the database foreseen and the motivation of its development. The way to achieve this goal is however quite long. If the production of a colour-magnitude diagram seems rather easy, the real situation is unfortunately not simple and the data are not always of sufficient quality. Therefore a large amount of work is necessary to plot a reliable colour-magnitude diagram. The various steps involved are:

1. the census of the stars in the cluster field, to determine the completeness in terms of both limiting magnitude and surface coverage. The database cross-reference tables and the collection of rectangular positions are useful in this respect;

2. the comparison of the various sources of data. The database provides tools to perform it and to compute mean values;

3. the selection of cluster members. This is certainly the main problem and no straightforward solution has yet been found. The other data contained in the database, like spectral types, radial velocities, remarks, and so on, may help assigning the membership;

4. with the best selected data, it is eventually possible to plot a nice colour-magnitude diagram and determine more accurate cluster distances and ages.

From that stage on, it becomes possible to do some astrophysical research. The results often depend on the reliability of the distance and age determinations. The other information collected in the database may be used to study the clusters' stellar content and properties. In spite of the limitations due to the lower precision of some data, or their limited number, the database is the best starting point for many astrophysical studies involving open clusters. Nowhere are complete data collections to be found and one merit of the database is to give a clear report of the present observation status.

The database facilities have been used by Meynet et al. (1993) to compare new theoretical isochrones with the colour-magnitude diagrams of 30 clusters. In the spring of 1994, the database was successfully used to cross-identify the stars detected by the Tycho experiment on board the Hipparcos satellite.

12–10   Future plans

The challenge for the future lies in the management of the rapidly growing number of new data coming from CCD photometry and extensive observations of faint stars in nearby clusters and covering a wide range of ground-based and space techniques. The structure of the database can accept any new data type even for one cluster only.

Due to the increasing quantity of available data, any study will become more and more time-consuming. One has therefore to think to more automated methods that rely not only on extensive data collections, but also on the transfer of knowledge to the computer, to let it do much of the work. Therefore, aside from continuing to collect and install new data and take other data types into consideration, the main development should consist of implementing more analysis facilities and calibrations. Suggestions and collaborations are welcomed.

An expert system, adapted from Jonathan (Frot 1988), has been implemented in the database with a number of rules that should help the user to determine if a given star is a cluster member or not. Rules have to be improved to resolve the ambiguities resulting from the curvature of the sequences in some photometric diagrams and to extend the number of data taken into consideration. Presently, it looks in the database for the UBV data, spectral types, proper motion membership probability and radial velocity, and distance from the cluster center and makes inference on the membership. The expert system does the work of extracting the data itself and proposes a decision according to the data it has found and the rules it knows to analyse the situation. This behaviour reproduces the way the same question is solved by a human user. It could of course use more information than it does now, speak English instead of French, the language in which it has been developed, and above all, have better rules. It would however be extremely useful to be able to sort out the member stars more or less automatically, on the basis of objective criteria.

Another important point would be to include in the database the knowledge contained not only in the data or procedures themselves, but in the article and review texts. Some could of course be recovered by searching the bibliography, but it may take a long time because the number of papers written every year on open clusters is increasing regularly and reaches the level of 200 papers for the best years. Therefore the best results or exciting hypothesis should be extracted from the papers and stored in the database. This is actually possible thanks to the hypertext facility offered by NCSA Mosaic or the WAIS text indexing and search facilities.

References

[1] Alter G., Ruprecht J., Vanysek V. (1970), Catalogue of Star Clusters and Associations. Akademiai Kiado, Budapest.
[2] Frot P. (1988), 3 systèmes experts en Turbo C (Sybex, Paris).
[3] Ianna P.A., Adler D.S., Faudree E.F. (1987), AJ **93**, 347.
[4] Johnson H.L., Morgan W.W. (1953), ApJ **117**, 313.
[5] Lyngå G. (1987), Catalogue of open clusters parameters (5th ed.), CDS.
[6] Mermilliod J.-C. (1976), A&AS **24**, 156.
[7] Mermilliod J.-C. (1986), Bull. Inform. CDS **31**, 175.
[8] Mermilliod J.-C. (1988a), Bull. Inform. CDS **35**, 77.
[9] Mermilliod J.-C. (1988b), in Astronomy from Large Database I, Eds F. Murtagh & A. Heck, ESO Conf. and Work. Proc. no 28, p. 419.
[10] Mermilliod J.-C. (1992a), Bull. Inform. CDS **40**, 115.
[11] Mermilliod J.-C. (1992b), in Astronomy from Large Database II, Eds A. Heck & F. Murtagh, ESO Conf. and Work. Proc. no 43, 373.
[12] Meynet G., Mermillod J.-C., Maeder A. (1993), A&AS **98**, 477.
[13] Press W.H., Flannery B.P., Teukolsky S.A., Vetterling W.T. (1986), "Numerical Recipes" (Cambridge Univ. Press, Cambridge).

[14] Ruprecht J., Balasz B., White R.E. (1981), Catalogue of Star Clusters and Associations, Supplement I. Ed. B. Balasz, Akademiai Kiado, Budapest.

[15] Schaller G., Schaerer D., Meynet G., Maeder A. (1992), A&A **96**, 269.

[16] van Leeuwen F. (1985), in IAU Symp. no 113, Eds J. Goodman & P. Hut (Reidel, Dordrecht) p. 579.

13

The HEASARC facility

The HEASARC team

13–1   HEASARC Overview

The purpose of the HEASARC ([→1]) is to support a multi-mission archive facility in high energy astrophysics for scientists all over the world. Data from space-borne instruments on spacecraft, such as ROSAT, ASCA (formerly Astro-D), GRO (Compton), BBXRT, HEAO 1, HEAO 2 (Einstein), EXOSAT, and XTE are provided, along with a knowledgeable science-user support staff and tools to analyze multiple datasets. The HEASARC activity is a joint effort between the Laboratory for High Energy Astrophysics (LHEA) and the National Space Science Data Center (NSSDC).

The LHEA is responsible for the individual science user support. This function provides access to an expert staff of astrophysicists cognizant of all the mission data that is supported within the archive. LHEA maintains an active database, the HEASARC On-line Service, for rapid access to and analysis of high energy astrophysics data from multiple missions. In addition, the HEASARC science support staff develops software analysis tools that facilitate the comparison of high energy astrophysics data in the archive. These activities are invaluable to researchers who are not knowledgeable about the appropriate use and analysis of high energy astrophysics data. Providing non-experts such ease-of-access to disparate datasets is an important function of the HEASARC. The LHEA also is restoring older data sets by reformatting them to FITS format. As part of this effort, the FITSIO and FTOOLS library of software has been developed to handle FITS files to be written, read, and manipulated.

The role of the NSSDC in the HEASARC is to maintain and provide a more routine access to the extensive physical archive that is being accumulated. This function includes bulk data distribution and distribution to individual scientists who do not need the extensive user support contacts that are provided by the LHEA/HEASARC staff. The NSSDC also works with scientists through the NSSDC/FITS office to meet their formatting needs. Additional NSSDC facilities are used to produce and distribute HEASARC CD-ROMs.

D. Egret and M. A. Albrecht (eds.), Information & On-Line Data in Astronomy, 139–146.

13–2 Motivation and Requirements

2.1 MOTIVATIONS

There are four distinct categories of motivations for archiving high energy astrophysics data: historical studies, theoretical follow-up, surveys and assurance.

2.1.1 Historical studies are the most obvious archival activity. An observer discovers a new phenomenon, or is studying one previously known, and needs to check earlier data to independently confirm its existence and/or track its long-term variability. These studies can be the most difficult type of archival activity because they involve combining and/or comparing datasets from different telescopes. The major issues to consider for historical studies are ease of access to data, and the ability to perform cross-instrument calibration. An activity related to historical studies is the use of archival data as part of a justification to propose the use of a new telescope.

2.1.2 Theoretical follow-up is the need to test new models against existing data. In many cases, the interpretation of a phenomenon can take many years, with theoreticians repeatedly building models and testing them against the data. Theoreticians previously had to work closely with the original investigator to test their models, or they had to make "eye-ball" fits to published data. The major issue for theoretical studies then is that the theoretician does not have a detailed knowledge of the instrument characteristics or analysis techniques. He or she simply wants a data product and the associated calibration to test against the model in a clearly-described, easy-to-read data format.

2.1.3 Surveys provide the opportunity to combine many observations of a single class of object (*e.g.* AGN) made by many different investigators using the same telescope and instrument. The current principal investigator approach to allocating observation time means that large, uniform samples of particular object types are rarely available to a single observer. Only after the data enter the public domain can a survey of the properties of a particular class of object be made. The main issue regarding surveys that the HEASARC faces is to ensure that a user can access a sample of all objects of a particular class.

2.1.4 Assurance is the ability to guarantee both that an observation is analyzed (and, if appropriate, published), and that unjustified repeat observations are not made. Observation time on satellites is very limited (and expensive). Making the data available after some fixed amount of time ensures that all interested parties in the field get access to that data. It also ensures that the data are eventually looked at. The issue here is that in many cases an observation may stay unpublished because the result is not sufficiently noteworthy. It is essential to provide a simple overview of the main results of the observation to avoid unnecessary repeated analysis of the raw data.

2.2 REQUIREMENTS

The four categories of motivations described above and the issues related to them place the following requirements on the HEASARC:

 ▷ the ability to perform multi-mission analysis

 ▷ a hierarchical archive structure

 ▷ the ability to perform a "quick-look" assessment of the value of data

 ▷ vendor-independent data formats.

13–3   Analysis of HEASARC Data

3.1 THE HEASARC DILEMMA

For every mission, data flow is identical. First, the raw data undergo some form of *data reduction* to produce *data products* –typically a photon list, an image, a spectrum, and/or a lightcurve. These products then are *analyzed* to produce some *results*, which then are (one hopes) *published*. Although the sequence of events is much the same for each mission, the dilemma facing the HEASARC is that every mission to date has produced a data set in a different format, with a different set of analysis software. This diversity of data formats makes the long-term support and distribution of a multi-mission archive problematic because every mission is a special case. In addition, combining datasets from different missions is non-trivial.

Mission-specific formats tend to be used throughout the data-processing chain. In many cases, the raw telemetry data involve pre-processing and packing of the data on the spacecraft so as to maximize the information transmitted to the ground. There may be multiple telemetry and onboard computer modes that can add to the complexity of reducing the data. The data products produced by the data reduction software are more generic. For example, a lightcurve is a time and a count rate. However, even for data products, each mission typically generates its own format. A notable exception to this rule is that images recently have begun to be distributed in FITS format. The results of each also sometimes are kept in mission-specific or vendor-dependent formats, as, for example, is an INGRES DBMS table.

The data access to mission data therefore is limited to a data-processing system produced by the project. These systems tend to be monolithic systems that are not optimal for long-term maintenance or general distribution and use by the scientific community.

Specific problems with project-generated data processing systems are as follows:

▷ They are custom-built for each mission, even though the underlying functions are the same.

▷ There is a failure to modularize and isolate the mission-dependent functions.

▷ Calibrations and methodology are embedded in the code.

▷ The code is vendor-dependent (*e.g.* operating system, compiler, DBMS).

The last point in the preceding list is particularly problematic. In the long term, it makes maintenance of the data-processing system difficult. The code must be ported repeatedly to new hardware and software platforms as technology evolves. With so many different missions in the HEASARC archive, this activity could be a never-ending and expensive task.

In addition, the user community is becoming increasingly demanding. Users require access to the original raw data, and they want to reduce it from within a familiar analysis environment, such as IRAF, IDL, or XANADU.

3.2   THE HEASARC SOLUTION

The root of the problem is that each mission produces data in different formats. Many of the data reduction and analysis functions are basically the same: the driving factor is decoding the different data telemetry and any mission-specific data product formats. Up to now, there has been little, if any, re-use of software between missions. The HEASARC solution is to reformat all data to a single, standard structure. This structure should be self-describing, so that the user need only look in the header to be able to read the file. The FITS standard provides such a capability.

FITS (Wells et al. 1981; Grosbøl 1991) is an International Astronomical Union (IAU) and NASA standard for distributing data analysis software. Moreover, there are FITS readers within all of the popular environments (*e.g.* IDL, IRAF, MIDAS). The adoption of the *binary table FITS* standard, which allows the byte structure of each column to be defined in the header, was a real breakthrough. This standard allows compact table structures to be defined that can mirror the underlying table structures in most data analysis systems such as MIDAS or IRAF STSDAS tables.

The HEASARC distributes all *useful* data as FITS binary tables, including telemetry. Although at first sight it may seem a formidable problem to reformat a complex telemetry stream containing science and housekeeping data, it actually is simpler than having to build from scratch a data reduction and analysis system. Reformatting the data forces an isolation of the mission-specific function of decoding the telemetry. The following data reduction tasks have both mission–specific and mission–independent functions. By reformatting the telemetry, it is simple to recycle the mission-independent functions.

To implement this plan, the HEASARC took the following steps. First, the data reduction system for the ASCA (formerly Astro-D) mission were constructed so as to form the basis for a multi-mission infrastructure. The ASCA telemetry was reformatted to FITS, and all of the mission-dependent and independent bits were isolated. Second, the HEASARC reformatted existing telemetry and data products from past missions, such as Einstein, HEAO 1, and EXOSAT. The experience learned and the FITS file structures defined then could be fed into future missions such as XTE.

To enable both the HEASARC and future missions to reformat to FITS, the HEASARC provides a portable FORTRAN 77 subroutine library to write and read FITS files. The package, known as FITSIO, was released in early 1992. The HEASARC also has defined mission-independent FITS file structures for spectra, lightcurves, and photon lists. These structures allow data products to be distributed transparently between different analysis packages. In particular, the HEASARC and the ROSAT Data Center have defined a set of "rationalized" FITS files for the ROSAT archive. These rationalized files differ from previous files in that the structure and keywords have a multi-mission flavor.

The HEASARC does not force the community to use one data analysis environment. Instead, it has a policy of ensuring that any HEASARC-produced data reduction tasks are distributed in ANSI standard code, with the input and output only operating on FITS files. In addition, all parameter checking and binding is isolated, so that these packages can be interfaced to the user's favorite analysis environment.

To facilitate the HEASARC's approach, a Data Selector was produced in collaboration with the ASCA mission to allow Boolean selections from FITS tables. This data selector forms the basis of a multi-mission data reduction system, and it is very similar in concept to the MIDAS and STSDAS table systems. The major advantage of the HEASARC data selector is that it operates directly on FITS tables, making the system fully portable. It is written strictly in FORTRAN 77. The software isolates the parameter input and validation from the kernel that actually does the task, thus allowing the selector to run under different analysis environments. The first version was built to run under both the IRAF (using the FORTRAN interface), and the XANADU environments. It is trivial for other developers to integrate the selector to their own analysis environments, so long as their environments have an isolated parameter interface.

The remaining mission-dependent part of any data reduction system is the calibration data. The HEASARC has defined standard formats for distributing calibrations. Like the data themselves, calibrations can be divided into raw data, such as detector energy resolution function; or a telescope point spread function, such as a detector response matrix or an exposure map. The HEASARC encourages any future developers to externally define all calibration information so that it can be accessed by any data reduction or analysis system.

13–4   Distribution of HEASARC Data

Distribution of HEASARC data is performed via on-line access and by mass distribution methods such as CD-ROM. The data on CD-ROMs are distributed on a regular basis. These CD-ROMs contain primarily data products and catalogs from each mission. In addition to the CD-ROMS, data can be accessed remotely via on-line services.

On-line services such as SIMBAD, NED, IUE, EXOSAT, and Einline are well known and work well at delivering the data quickly to the user. The disadvantage to these various services is that each one has a different user interface with which the user must become familiar. NASA's Astrophysics Data System (ADS) uses a client-server approach to allow remote queries of databases. The archive sites retain control of the archive contents, but rely on a common user interface provided by the central organization.

In addition to ADS, the HEASARC provides an on-line service to allow remote login to the HEASARC data holding and to data analysis software. The emphasis is on *browsing* of the data, such that a user can make a quick-look assessment of its worth before exporting it —or part of it— to his or her home site. Rather than invent yet another on-line system, the HEASARC adopted an existing system—the one developed for the EXOSAT mission by the European Space Agency (ESA). The advantage of this system is that it provides the capability to not only access the data, but to display and analyze it remotely.

At the heart of the system is the *Browse* program, a command-driven environment that allows a user to search one or more database tables by coordinates, name, object class, or any other valid parameter combination. The user then can display the selected data, or run analysis software on it.

13–5 Introduction to SkyView

SkyView is a Virtual Observatory on the network. Astronomers can generate images of any portion of the sky at wavelengths in all regimes from radio to gamma-ray. Users tell *SkyView* the position, scale and orientation desired, and *SkyView* gives users an image made to their specification. The user needs not worry about transforming between equinoxes or coordinate systems, mosaicking submaps, rotating the image, *etc.*, *SkyView* handles these geometric issues and lets the user get started on astronomy.

SkyView is available on the World-Wide Web at [→2] and also through telnet. *SkyView* has two interfaces: the web interface and the interactive interface. The web interface uses HTML forms to format a request and sends the request to *SkyView*. A page with a GIF of the region requested is displayed with anchors that retrieve a FITS image or images is returned to the user.

Two HTML forms are available. The basic form allows the user choice of surveys, area and size of image, but is simple enough to fit in a default-size Mosaic window. The advanced form gives the user more detailed control over the size and resolution of the image, and allows combining data from multiple images and overlaying images with catalog markers.. With these two forms, the labels for each of the fields are anchors to documentation on the meaning and use of the field.

The Interactive interface lets the user work interactively with images. It starts a remote X-windows display on the user's host machine which runs a *SkyView* session on the *SkyView* host. Users have all of the capabilities of the web interface but can also manipulate the images they retrieve. Users can play with the color tables, smooth images, look at catalogs interactively, get the coordinates of points within an image, and manipulate contour and image overlays.

Users can get the interface directly from the web, or through the generic xray account set up by the HEASARC.

Users log in as xray on legacy.gsfc.nasa.gov. This account has no password and allows users to access a wide variety of HEASARC services. The skyview command in this environment starts and interactive *SkyView* session.

Further information is available in the "*SkyView* Users Guide" which is available on the net ([→2]) and through anonymous ftp ([→3]) Information of the data available in *SkyView* is available in "The *SkyView* Surveys Guide".

NOTE FROM THE EDITORS

This chapter has been compiled from the on-line WWW service [→1] by the Editors.

References

[1] Grosbøl, P. (1991), in *Databases and On-line Data in Astronomy*, M. A. Albrecht & D. Egret (Eds.), Kluwer Acad. Publ., 253.

[2] Wells D. C., Greisen E. W., Harten R. H. (1981), Astron. Astrophys. Suppl. **44**, 363.

Access Pointers

[→1] HESARC: `http://heasarc.gsfc.nasa.gov/`

[→2] SkyView: `http://skyview.gsfc.nasa.gov/skyview.html`

[→3] SkyView Users Guide: `ftp://skyview.gsfc.nasa.gov.`

14

IPAC Datasets and On-Line Services

R. Ebert, G. Helou, and J. Mazzarella for the IPAC Tools Group

14–1   Context

The Infrared Processing and Analysis Center (IPAC) was created by NASA
to process and analyze data from the Infrared Astronomical Satellite (IRAS)
in 1983. Now that the IRAS work is completed, IPAC exists to perform
data processing and science support activities that are critical to other NASA
infrared astronomy missions. But the IRAS archive lives on, and continues
to be used extensively, both for ongoing research and to prepare for future
missions.

Riding the trend of faster networks and computers, IPAC has reinvented
most of its data access and processing services focusing on reaching users
over the network. The challenge is to provide the remote, on-line user with
useful access to the IRAS archive data while providing the expertise IPAC
users have come to rely on.

The cornerstone to the IPAC archive is the IRAS mission data. IRAS was
a joint project involving the United States, the Netherlands, and the United
Kingdom, that surveyed 96% of the sky in four wavelength bands (12, 25, 60
& 100 μm) in 1983. IRAS returned about 24 GB of raw data which produced
an archive of 132 GB when combined with time, pointing, and calibration
information. From these data IPAC has produced two types of products: (1)
The 7 major IRAS catalogs (see Table 14–1) contain nearly 1 GB of information
including source associations with other major astronomical catalogs. (2) The
IRAS Sky Survey Atlas (ISSA) contains over 4 GB of all–sky images which
have been flux and position calibrated (1.5′ pixels with 4′ resolution).

The IPAC on-line services are intended first to provide access to the IRAS
archive (catalogs, images, and raw scan data), second to support observation
planning and data analysis for new IR missions, in particular the European
Space Agency's (ESA) Infrared Space Observatory (ISO), and NASA's Wide–
Field Infrared Explorer (WIRE). Our goal is to go beyond the mechanical
access to databases and provide helpful and informative interfaces that act as

147

D. Egret and M. A. Albrecht (eds.), Information & On-Line Data in Astronomy, 147–152.
© 1995 *Kluwer Academic Publishers.*

interpreters between astronomer and computer system. These are interfaces that incorporate as much as possible of the scientific context and expertise already gained by IPAC while working with the data and their users.

14–2   Tools and Services

IPAC services are available from Internet hosts running an X Window System compatible server. The services run on IPAC computers and rely on the user's workstation for display. A few of these services are described here. A more complete list of IPAC on-line services is given in [→1] to [→8].

2.1   IRSKY

IRSKY is the primary IPAC software tool for viewing the infrared sky, offering convenient and efficient access to the major released science products from the IRAS mission. It is designed as an environment for astronomers to plan observations in the context of the known infrared sky, with special emphasis on observations using ESA's Infrared Space Observatory (ISO).

Users need only specify an object name or position on the sky, and the interface will display:

  ▷ Images from the all-sky IRAS Survey Atlas, with a resolution of 4′ per pixel in each of the IRAS wavelength bands at 12, 25, 60, and 100 μm, with a model of the zodiacal emission subtracted (the interface allows user control of image display parameters);

  ▷ Markers overlaid on the images showing the location of entries in the IRAS Point Source and Faint Source Catalogs, as well as the Hubble Space Telescope Guide Star Catalog, and the ISO Guaranteed Time Observations list (catalog information can be displayed by simply clicking the mouse on a marker);

  ▷ Estimates of the total sky brightness at any wavelength between 5 and 200 μm, and estimates of confusion noise due to source crowding and emission structure.

IRSKY provides a summary guide to the ISO instrument parameters, and a graphic representation of the ISO focal plane (with apertures, chopping directions and Sun vector) and of ISO mapping rasters. Additional data sets and functions will be added to IRSKY in the future, based mostly on user demand.

Access to the IRSKY service does not require a login name or password (see [→1]). The interface has an internal help system of files and Motif style context-sensitive help. To further guide the user in interpreting results, separate documentation on IRSKY is available for viewing independent of the interface (see [→2]).

There is also an electronic mail interface to IRSKY that provides access to the background estimators and catalog search functions. The IRSKY Batch Inquiry System (IBIS) instructions are available on-line [→3].

2.2 XCATSCAN

XCATSCAN [→4] is an interactive software tool for scanning IRAS databases and other catalogs, and extracting parameters for sources which meet specified criteria. The point-and-click user interface is available for remote access over the Internet from computers running X, and it is designed especially for scientists who wish to make queries of IRAS databases in conjunction with other major astronomical catalogs. Catalogs may be queried by specifying location on the sky (position and search radius, coordinate "boxes," or IRAS source name), constraints on data fields (*e.g.* $f_\nu(60\mu m) > 10.5$), and positional associations between IRAS sources and objects in other catalogs. Catalog columns may be easily queried by filling in a table, or by direct entry of constraints in Structured Query Language (SQL). The results may be saved as ASCII or FITS tables and transferred via FTP to your computer for analysis.

Recent features include the ability to upload coordinate lists (private catalogs) via anonymous FTP and scan the on-line catalogs for positional or relational matches, the ability to utilize the positional uncertainty ellipses in the IRAS catalogs to estimate the degree of overlap with input sources, and access to the new IRAS Optical Identification (OPTID) catalogs. A complete list of available catalogs is given in Table 14–1.

2.3 XSCANPI

XSCANPI [→5] is an interactive software tool for viewing, plotting and averaging the calibrated survey scans from IRAS; these scans are the fundamental data from the IRAS survey. XSCANPI is useful for measuring the fluxes of extended, confused or faint sources, for diagnosing source extent, and for estimating local upper limits at an arbitrary position on the sky. The sensitivity is comparable to that obtained in the IRAS Faint Source Survey—about a factor of 2-5 deeper the IRAS Point Source Catalog (PSC), depending on the local noise and number of scans crossing the target position.

XSCANPI allows you to interactively subset, plot and coadd IRAS scan data at any sky position of interest from any workstation connected to the Internet. Source lists can also be uploaded via anonymous FTP and read by XSCANPI. Due to the compute-intensive nature of this processing and

IPAC On-Line Catalogs
IRAS Faint Source Catalog, v. 2.0[a]
IRAS Point Source Catalog, v. 2.1[b]
IRAS Cataloged Galaxies and Quasars
IRAS Optical Identifications (OPTID)
IRAS Serendipitous Survey Catalog
IRAS Small Scale Structure Catalog
IRAS Pointed Observation Products Catalog
Fourth Cambridge Radio Survey Catalogue
Bright Star Catalog, 5th Ed. (prelim)
Catalog of Infrared Observations, 3rd Ed. - March 94
Catalog of Nearby Stars, 3rd ed. (preliminary)
Dearborn Observatory Cat. of Faint Red Stars
Fifth Fundamental Cat., Basic Fundamental Stars
General Catalogue of Variable Stars, 4th Ed.
ISO GTO Target List
Johnson UBVRI Photometric Catalog
NGC2000.0 by J.L.E. Dreyer
Photometric Data for the Nearby Stars (and Systems)
Catalog of Positions and Proper Motions - PPM (N, S, Supp) [c]
Veron and Veron-Ceti Catalog of Quasars & AGN
Revised AFGL IR Sky Survey, Primary Data
Second Reference Catalog of Bright Galaxies
CFA Redshift Catalogue, Main Data File
RNGC main data file
SAO J2000 Star Catalog
Seyfert Galaxies, data & references
Two-Micron Sky Survey, data ordered by RA
Catalog of Homogeneous Measurements in UBV Systems
Uppsala General Catalogue of Galaxies
Yale Zone Catalog (in multiple tables)

[a]Includes positional associations and table of Rejects.
[b]Includes positional associations and tables of Rejects, HCON and WSDB information.
[c]Contains all records but only a subset of fields (columns).

Table 14–1: IPAC on-line astronomical catalogs (Nov. 1994)

limited resources, access is limited to a small number of concurrent users. This restriction may be lifted in the future. XSCANPI is maintained as a remote-access tool at IPAC instead of being exported product because it relies on the IRAS Level 1 Archive, which even after a factor of 5 compression takes up ~ 25 GB on an optical disk jukebox. XSCANPI is intended for immediate access to the basic IRAS data for users wishing to process a few positions interactively. For batch processing of large source lists, the batch SCANPI software which has been available at IPAC for many years is still available.

2.4   WWW

IPAC is also making considerable use of the World-Wide Web (WWW) for distribution and on-line access to IPAC documentation and services [→6]. One example of unique WWW services at IPAC is the ISSA Postage Stamp Server [→7] which in turn uses other IPAC servers and the CDS SIMBAD service to deliver IRAS images for a given object name or celestial position.

Distributed information systems like the WWW and tools like the National Center for Supercomputing Applications (NCSA) Mosaic have changed the way people navigate the network, and their expectations for easy-to-use, intuitive access to on-line services. IPAC will continue to look at the emerging standards and tools in this arena to offer its users a combination of new HTML-Forms based interfaces as well as the more interactive and application specific X/Motif based interfaces.

14–3   Future Plans

As the IRAS data products reach their tenth anniversary, IPAC is turning its attention to supporting large new missions. The largest of these will be the Two Micron All-Sky Survey (2MASS) which as currently planned, will survey the sky from the ground in three near–infrared wavelength bands (J, H, K) with 2″ pixels. Beginning in 1996 and lasting 2 to 3 years, this project will return nearly 10 TB of data, at the rate of 10 GB per night, or the equivalent of an IRAS mission every 3 days. The processing is expected to produce over 4 TB of images, and 2 GB of catalog information containing 10^8 stars and 10^6 galaxies. Preparation and prototype observations for this project are already under way. Although the data processing requirements can be met with currently available processors, much will be gained from improvements in data handling and storage technology for providing access to terabyte archives. Ideally, 2MASS data could go on-line as soon as they are validated.

IPAC is also the US science support center for European Space Agency's Infrared Space Observatory (ISO). The ISO mission is expected to return 350

GB of data, 30 – 60 GB of which will be for US principal investigators. In the coming year, IPAC will provide ISO science and instrument expertise, and software to assist the US community with proposal preparation, and to help the US Guaranteed Time Observers (GTO) and Guest Observers (GO) with detailed observation planning. During and after the mission, IPAC will support the community with detailed analyses of the ISO archive. In keeping with current trends, most observers may want to analyze their data without physically traveling to IPAC. This will challenge IPAC to find ways to render expertise and support conveniently over the networks to individual researchers as well as anonymous users.

14–4   Conclusion

IPAC's vitality derives from the synergy of its three main goals: produce high quality data products, develop expertise in those data products, and provide the astronomical science community with access to those products and expertise. The success of our efforts to use the Network to provide on-line access depends on our ability to make our services functional and friendly to novice and expert user alike, even as the volume of data and information increases 1000-fold in the next five years.

Acknowledgements

All members of the IPAC Tools Group are contributing to the software and systems discussed here. They are J. Bennett, B. Hartley, L. Hermans, I. Khan, M. Kong, G. Laughlin, S. Lord, R. Narron, and D. Van Buren.

IPAC is operated by the Jet Propulsion Laboratory, California Institute of Technology for the Astrophysics division of NASA.

Access Pointers

[→1] IRSKY service: `telnet://irsky.ipac.caltech.edu:1040/`
[→2] IRSKY information:
        `http://www.ipac.caltech.edu/ipac/services/irsky.html`
[→3] IBIS: `http://www.ipac.caltech.edu/ipac/services/ibis/ibis.html`
[→4] XCatScan: `telnet://xcatscan@xcatscan.ipac.caltech.edu/`
[→5] XScanpi: `telnet://xscanpi@xscanpi.ipac.caltech.edu/`
[→6] IPAC Homepage: `http://www.ipac.caltech.edu/`
[→7] ISSA-PS: `http://www.ipac.caltech.edu/ipac/services.html#issaps`
[→8] IPAC Service Directory: `http://www.ipac.caltech.edu/services.html`

15

The Archives of the Canadian Astronomy Data Centre

Dennis R. Crabtree, Daniel Durand, Severin Gaudet,
Norman Hill & Stephen C. Morris

15–1 Introduction

In 1986, with the launch of the Hubble Space Telescope imminent, the Dominion Astrophysical Observatory (DAO) established an agreement with the Space Telescope Science Institute (STScI) whereby the DAO would receive a copy of HST data and serve as the Canadian archive centre . The Canadian Astronomy Data Centre (CADC) was the group which was formed to manage the Canadian HST archive.

Besides serving as the Canadian HST archive centre, the CADC is also responsible for archiving data from the Canada France Hawaii Telescope (CFHT). This is one of the few archives of data from large, ground-based telescopes which currently exists. The design and implementation of the CFHT archive was based upon the experience we had gained from establishing the HST archive in Canada. The CADC is fairly unique in that we have archives of both ground- and space-based data, each of which presents its own opportunities and problems.

One of the main goals of the CADC is to develop tools and techniques which increase the scientific usefulness of the archives or which provide more efficient access to the archives. We have added our own *features* to the HST archive and are continually working on ways to make the CFHT archive more scientifically useful.

Both the CFHT and HST archives are accessed via STARCAT which was developed by the ST-ECF, ESO and STScI to access the interim HST archive known as the Data Management Facility (DMF). At ST ScI the DMF and STARCAT will soon be replaced by the Data Archive and Distribution System (DADS). However, as we needed an interface for both the CFHT and HST archives, the ST-ECF planned on maintaining STARCAT and ESO needed an interface for the planned VLT archive, we all decided to continue develop-

D. Egret and M. A. Albrecht (eds.), Information & On-Line Data in Astronomy, 153–162.

ment of STARCAT and to expand its capabilities to handle multiple archives. STARCAT also offers access to a number of other astronomical catalogues, some of which contain pointers to archival data such as IRAS Low Resolution Spectra or IUE spectra.

Information on the CADC and the services it offers is available at [→1].

15–2   HST Archive

The initial expectation was that CADC would simply use the same software and hardware in use at ST ScI so that running the HST archive would be relatively easy. Due to a variety of problems getting the ST ScI software running, and a desire to move away from the Vax/VMS operating system, we migrated the HST archive to a Unix-based system in collaboration with the Space Telescope – European Coordinating Facility (ST-ECF). This collaboration has been very successful and our groups have been much more productive than either group would have been on its own.

We maintain a copy of the HST database on a SUN/Sparcstation 2 running the commercial database system Sybase, which is the same database system used at ST ScI. In the past we kept our copy of the database current with respect to the master copy using a software tool we developed named DBsync. With the move to the permanent archive system DADS at ST ScI, this mechanism has been phased out in favour of Sybase's *replication server* which offers the same functionality. Our copy of the HST database is updated immediately when records are added to the master copy in Baltimore or when any changes are made.

We receive a copy of HST data immediately after it has been archived by the DADS system on 12-inch Sony WORM optical disk. These disks have a capacity of 6.5 GB. We receive both raw and calibrated data as well as all calibration data. This, and the fact that we have weekly updates of the Space Telescope Scientific Data Analysis System (STSDAS), gives us the ability to reprocess data as newer calibration data or procedures become available.

Users access the HST archive using the STARCAT software which is described in more detail below. Briefly, STARCAT provides a screen-oriented interface to the archive and allows the user to specify qualifications on any of the fields available on the screen. As the Hubble catalogue is very large there are many screens available, each of which presents somewhat different information. One way in which we have increased the scientific usefulness of the Hubble archive is to produce a new database table which summarizes information in the database on *scientific targets*. To view this table we have developed a *science screen* (see figure 15–1). This screen displays only observations of scientific targets and gives scientifically relevant information such as exposure time, filter, spectral region, *etc.*

15–3   CFHT Archive

Data taken at most ground-based observatories is not currently archived. We have used the expertise and experience gained by working with the HST archive and applied it to producing an archive of CFHT data. Most of the work on the CFHT archive has been done at the CADC but this project would not have been possible without the support of the staff at CFHT and the CFHT Board of Directors. Data from the CFHT is proprietary for a period of two years after which it is publically available and unlike the HST data we distribute data to the world astronomical community. The CFHT archive contains data from early 1991 onward which currently amounts to ≈ 120 GB.

The *data handling pipeline* at the CFHT has been configured so that every data frame taken at the summit is automatically and transparently copied to a staging area on magnetic disk at the headquarters building in Waimea. All of the data is stored in FITS format. There is a process on the archive computer in Waimea which periodically looks for new files in this staging area. When it finds new files it first makes a copy of the FITS header in a separate file. The process then copies the file to optical disk (Sony 12-inch WORM , capacity 6.5 GB), verifies the copy on the optical disk and then deletes the file from magnetic disk. The next day all of the FITS headers are sent electronically to the CADC while the optical disk is sent when it is full.

When the FITS headers from the previous night's observations are received they are automatically parsed and used to update a series of tables in the database. Unless there is a communications problem the CFHT observation catalog is never more than 24 hours out of date. When the full optical disks are received their contents are scanned and the table in the database which acts as a master directory is updated. A consistency check that a file exists for each entry in the database is also performed. Thus far we have not lost any data files.

As one of our efforts to increase the usefulness of the archive we are planning on passing the direct imaging data through an automatic processing pipeline which will produce bias subtracted and flat-fielded images (Crabtree *et al.* 1994). This process works by first querying the database to select appropriate files and then using IRAF to perform the actual processing. While the processing may not be the best that is possible, the data will be of high enough quality to be useful for certain applications. These processed images will also be used to generate *preview* images, in a similar manner to the HST images (see the following section). Our goal is to develop automatic processing pipelines for some of the other instruments but this effort will likely be limited because we do not always have reliable access to information which is needed to automatically calibrate the data.

15–4   Accessing the Archives: STARCAT

Access to the HST and CFHT archives is provided through the STARCAT software. While the original STARCAT was essentially an ASCII terminal-based interface, the current STARCAT has evolved into much more of an X Windows application. The core part of STARCAT is still terminal based but many of the new features are spawned off as X applications. STARCAT now works in *client-server* mode. Users run STARCAT on their local workstation and the connection to the database, either at the CADC or the ST-ECF is made over the Internet. This makes very efficient use of the network as only a minimum of information is transferred.

4.1   PREVIEW

A major innovation which we have developed within STARCAT is the ability to preview data in the archives before requesting the data. Since HST and CFHT images are about 10 MB in size, it is impractical to keep them all on-line. However, archival researchers would benefit from being able to visualize the data before requesting it from the archive. Ideally, they would be able to confirm that their object of interest was present, the exposure was long enough, and generally assess the quality of the data. This preview capability allows archival researchers to do all of the above.

The data for the *Preview* mechanism is stored within the relational database. This data is in a highly compressed format so it takes up much less space than the original data. In the case of HST WF/PC images a typical image is stored in $\approx$ 60 KB rather than original size of 10 MB. Briefly, this compression factor is achieved by converting the data to 2-byte integers, spatially sampling the data by a factor of two in each dimension and applying a lossy compression algorithm to the data.

When an archival researcher using STARCAT somewhere on the Internet requests a *preview* the compressed image is retrieved out of the database, transferred over the Internet to the users machine, decompressed and dis-played. Since the *compressed* image is sent over the network and the decom-pression is quite fast, the previewing process is fairly efficient. We use a standard tool, SAOimage, for displaying the image as this is familiar to most astronomers and saved us from developing yet another display tool.

We also stored compressed versions of the HST spectra on-line for preview purposes. In this case the compression is non-lossy so the preview data is the same as the actual data. We use another public domain software package named *xmgr* as the display tool for spectra.

4.2 REQTOOL

In order to make the selection and retrieval of data from the archives as easy as possible we have developed an X11 program named *Reqtool* with which the user manages his/her request for data. Reqtool is started as soon as the first dataset is requested and provides the user with a current view of the request (see figure 15–2). The user can select the media for the archive request and in the case of HST data can choose which of the individual files in the dataset are needed. It is also possible to request the calibration files needed to recalibrate the request science data.

4.3 SCIENTIFIC VIEWS OF THE ARCHIVES

The HST and CFHT catalogs contain fields for the object (target) which was observed. In the case of HST these targets follow the guidelines (for the most part) specified during the proposal submission. However, one finds observations of both NGC 1275 and 3C84 so in searching by target name one needs to *know* the various names of an object. In the case of the CFHT catalog this is a very big problem as the object names are entered by the observer at the telescope so there is no control over the object name entered. Besides the problem noted above for HST there are also problems with case, typos, and other nonsense entries. For example CFHT has observed objects (as named by the observer) such as: NGP2, NO 60, I 231.

To address this problem STARCAT allows one to search the catalogue using a specified radius around a particular position. One does not always have coordinates available or coordinates that in the same epoch as those of the catalogue. STARCAT allows for an *object-centered* search of the catalogues by taking the name of the object as entered by the archive user, connecting to the SIMBAD database to get the coordinates of the object in the proper epoch for the catalogue and using these coordinates for the search. One advantage of this approach is that the user can enter any of the names by which an object is known.

Another problem with observation catalogues is that there is very limited information, such as spectral type, redshift, etc., on the target. Thus if one wants to find archival data on quasars with redshifts between 0.3 and 0.6 with B-V colours > 0.3 one has to first compile a list of objects which meet those criteria and then search the archive catalogue for each of the objects.

STARCAT provides a dynamic mechanism, although yet limited at this time, to search the archival catalogues for data of a particular scientific interest. This is done by providing the user with a *screen* which provides access to both a scientific catalogue as well as the archive catalogue. Figure 15–3 shows an example of the screen which allows researchers to search the HST archive for data on quasars including the ability to qualify the search based upon the fields available in the quasar catalogue such as redshift.

We feel this is potentially a very powerful mechanism for searching the archives and we have produced screens which allow for searches of the HST and CFHT archives in combination with Hewitt-Burbidge Quasar Catalogue, the RC3 catalogues of galaxies and the SAO stellar catalogue.

15–5   Other Data

STARCAT also offers access to a variety of other astronomical catalogues and provides powerful relational database techniques for searching these catalogues. Two catalogues are worth mentioning in more detail. STARCAT gives access to the IUE Uniform Dispersion Low Dispersion (ULDA) archive produced by the IUE VILSPA group. We have loaded the ULDA data into the database and linked it with the IUE observation log via our data preview mechanism. This makes over 50,000 IUE low dispersion spectra available on-line. We also offer access to the ULDA through the software (USSP) distributed with the data but the STARCAT access provides the capability of viewing the spectra interactively.

In cooperation with the IRAS group at the University of Calgary we have loaded over 11,000 IRAS Low Resolution Spectra (LRS) into our database and linked them with the IRAS Point Source Catalog (PSC). This gives interactive access to the spectra combined with the ability to qualify searches of the PSC based upon flux levels, etc.

15–6   Future

The CADC is working with the James Clerk Maxwell Telescope (JCMT) on developing an archive of JCMT data at the CADC. Their data consists of 1-d spectra and fit very nicely our system for keep spectra on-line for interactive viewing. We expect this project to be complete in early 1995 and the first data to be available in late 1995 or early 1996.

The CADC expects to develop the archive for the Gemini Project which will see two 8-m telescopes contructed, one on Mauna Kea and one on Cerro Pachon in Chile, being shared by the US, UK, Canada, Chile, Argentina and Brazil.

References

[1] Crabtree, D.R, Irwin, A., Blaber, R., Gaudet, S., & Durand, D. (1994): in *Handling and Archiving Data from Ground-based Telescopes*, Eds. M. Albrecht & F. Pasian, ESO Conf. and Workshop Proceedings Series, **50**.

Access Pointers

[→1] CADC: `http://cadcwww.dao.nrc.ca/`

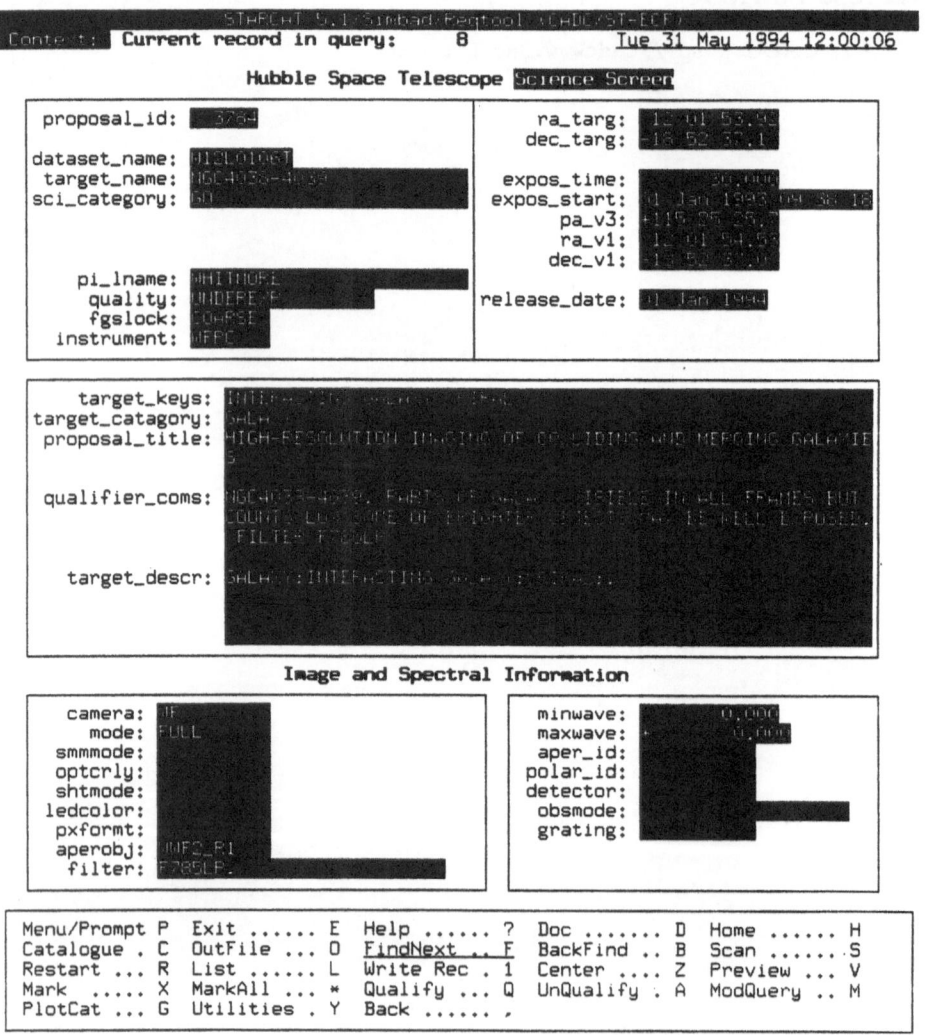

Figure 15–1: HST "science" screen in STARCAT

							Media		
File	Submit	Requests	Help				FTP	EXAB	DAT

Current requests:

Date Submitted	Archive	Num. Datasets	Num. Files	Size K. Bytes	Media	Status	Request Menu
current	hst	7	47	99851	DAT	Not submitted	

Datasets for archive hst current request:

Dataset Name	Target Name	Num. Files	Size K. Bytes	Status	Dataset Menu
W0PF0G01T	NGC596	7	15275		
W0PF0G02T	NGC596	7	15275		
W0PF0G03T	NGC596	7	15275		
X1AR5Q01T	UGC8745	5	8226		
W10A0D01T	NGC5322	7	15264		
W10A0D02T	NGC5322	7	15264		
W10A0D03T	NGC5322	7	15264		

Figure 15–2: STARCAT's ReqTool for managing data requests

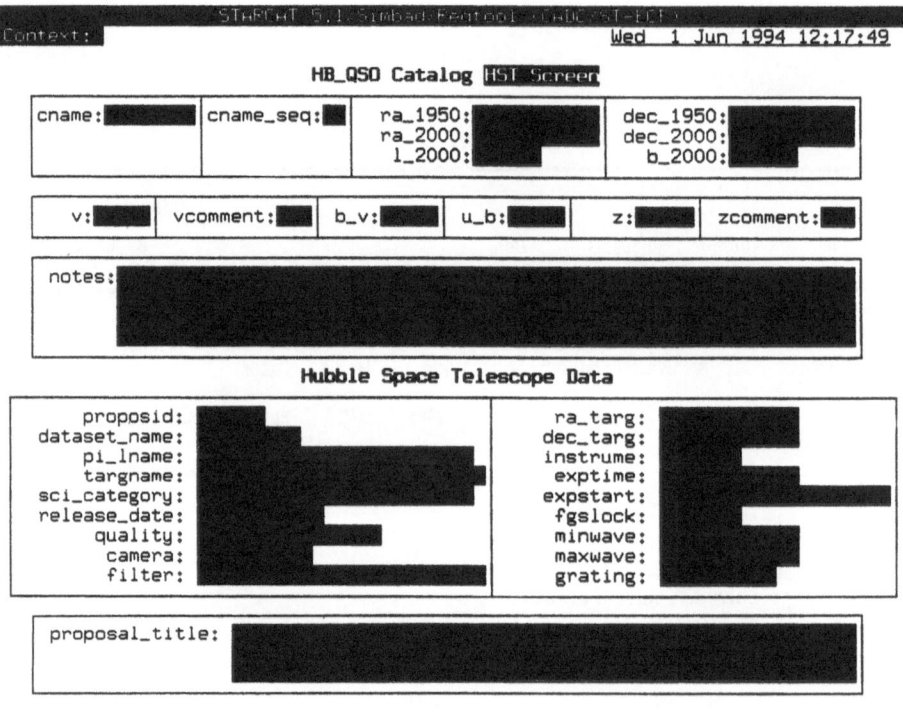

Figure 15–3: Searching the HST archive for Quasars in STARCAT

16

A global perspective on astronomical data and information: the Strasbourg astronomical Data Center (CDS)

D. Egret, M. Crézé, F. Bonnarel, P. Dubois, F. Genova, A. Heck, G. Jasniewicz, S. Lesteven, F. Ochsenbein & M. Wenger

16–1 Introduction

The Strasbourg astronomical Data Center (CDS) has been created in 1972 with the following goals:

▷ to collect useful data concerning astronomical objects, and available on electronic form (these are mainly observations produced by ground-based and space-borne observatories)

▷ to improve them through comparisons, critical evaluations, and cross-identifications

▷ to distribute the results to the astronomical community

▷ to conduct scientific research using these data.

In this chapter we describe how the CDS works at providing a global perspective on astronomical data and information, with the help of recent technological developments.

The CDS is a laboratory of the French 'Institut National des Sciences de l'Univers', an institute of the 'Centre National de la Recherche Scientifique (CNRS)'. It is installed at Strasbourg astronomical Observatory, depending also from Louis Pasteur University.

The increasing efficiency of cameras and other photon collecting devices used by astronomers, either from ground-based telescopes, or from space experiments, is generating an unprecedented accumulation of data. The ability of storing, managing, and giving access to this huge quantity of data, and the associated documents, is one of the major challenges of our science (and of natural sciences in general) for the next decade.

D. Egret and M. A. Albrecht (eds.), Information & On-Line Data in Astronomy, 163–174.
© 1995 *Kluwer Academic Publishers.*

The role of the data centers is to bridge the gaps between the special-ized approach of the scientific teams (where the detailed expertise about the specific data resides), and the general approach of the wider community of researchers (who need an easy access to data, calibrated as far as possible in meaningful physical units). A striking example of a new approach made possible by the data center activity is panchromatic astronomy (see Wells, 1992), a key to the understanding of many astrophysical processes, which implies to cross-compare data acquired for the same astronomical objects through several instruments working at different wavelengths.

The concept of data center has to be somehow extended in the context of the fastly growing number of on-line services of interest to astronomers, to help them retrieving the data and information they need among the many possible distributed sources.

16–2   The astronomical catalogues

The key role of CDS, as one of the first major astronomical data centers, is to collect the catalogues and act as a depository of catalogues and archives.

The early concept promoted by the CDS was to rely upon experts in order to collect catalogues —that is homogeneous sets of state of the art data, hopefully calibrated and in physical units.

Until recently, data center managers could easily identify individually authors who might be potential source of interesting data: they would contact those people and ask them to contribute to the general effort by providing a machine-readable catalogue. The data centre staff would then make the additional effort to properly edit, format and document the catalogues with their own standards. This process is no longer realistic due to the very fast development of digital data and electronic facilities.

In the recent past, data centers could also conceivably plan to store in magnetic tape collections whatever piece of information was produced elec-tronically in astronomy. Obviously the evolution of observing techniques, including modern telescopes with their digital receivers, radio telescopes, or space missions, now imposes a different approach: both storage capabilities and expertise on data cannot be concentrated at data centers anymore.

To face these evolutions, a number of initiatives, taken by the CDS, allow a more efficient approach of the collection and distribution of astronomical catalogues.

2.1 EXCHANGE AGREEMENTS

The CDS has signed international exchange agreements with NASA Astronomical Data Center (ADC), Japan's National Astronomical Observatory in Tokyo, the Russian Academy of Sciences, the SERC Starlink network in Great-Britain, China's Beijing Observatory, the University of Porto Allegre in Brazil, the University of La Plata in Argentina, and India's InterUniversity Center for Astronomy and Astrophysics.

2.2 STANDARDIZATION OF THE FORMAT AND DOCUMENTATION

Recently, a new standard for the format and documentation of ASCII tables has been proposed (Ochsenbein, 1994). The corresponding catalogue descriptions can be read easily by humans and by computers.

A full library of documented catalogues, available on-line for electronic distribution, is gradually built following this format. Procedures are also provided to translate this format into FITS, the common standard for images and binary data (Wells et al., 1981; Grosbøl, 1991).

The new standard is now also used by other data centers (NASA/ADC) and publishers (AAS CD-ROMs).

2.3 TABLES FROM ASTRONOMY & ASTROPHYSICS AND OTHER MAJOR JOURNALS

Some of the major astronomical journals are heading towards an electronic publication, at least of their data tables (see *e.g.* , for the AAS, Newsletter 62, October 1992).

Since January 1993, following an agreement with the Editors of Astronomy & Astrophysics, a number of tables from the main journal and from the Supplements are deposited at the CDS, and made available for electronic distribution. It is the Editor's initiative to decide whether data proposed by the authors as part of a paper should be printed only, or printed and made electronically available, or only provided in electronic form. Eventually the CDS staff is responsible for the integration of the corresponding files into the CDS standards.

The first issue of the AAS CD-ROM, in 1994, has also been put on-line at CDS, after an agreement with the Editors.

2.4 A CATALOGUE ALERTING SERVICE

New catalogues and tables made available on-line, and more generally catalogues for which an updated version has been produced in the current month are listed in a specific file on the ftp server. Users and data managers can suscribe to an automatic two-weekly alerting service through electronic mail.

2.5   A CATALOGUE SERVICE ON THE WORLD-WIDE WEB

The CDS started by 1991 an effort to move most frequently requested catalogues, together with their documentation, into a fully electronic archive accessible on-line. New catalogues and tables are now directly entered into this system. This effort was made possible through the collaboration between all data centers, and especially NASA/ADC which had in the meantime installed its own magnetic tape collection into a near-line system, and had produced a subset of most frequently used catalogues on CD-ROM.

The "Astronomer's Bazaar" made available at the end of 1993 (Egret and Ochsenbein, 1994) is a fully interactive on-line service, based on the World-Wide Web (WWW) allowing:

 ▷ to query the list of catalogues, by keyword, or in browse mode,

 ▷ to display the corresponding documentation,

 ▷ and to retrieve the complete electronic files (eventually compressed), from the anonymous ftp space of the CDS server.

More than 1000 catalogues and tables, for a total of several Gigabytes of data were already available through this procedure in 1994.

2.6   DICTIONARY OF THE NOMENCLATURE OF CELESTIAL OBJECTS

The Second Reference Dictionary of the Nomenclature of Celestial Objects has been recently published by Lortet *et al.* (1994). It can be queried through different keys (catalogue name, author, object type, format, *etc.*) on-line on the SIMBAD host, and through the World-Wide Web server. The Dictionary is also used as an auxiliary database in Simbad.

16–3   The SIMBAD database of astronomical objects

The SIMBAD database provides a unique and crucial view of the astronomical data: namely, organizing the information per astronomical object. This can only be done through a careful cross-identification of catalogues, lists, and journal articles, a task which has made SIMBAD a key tool used worldwide for all kinds of astronomical studies.

The SIMBAD database has been described by Egret, Wenger, and Dubois (1991) and recent developments are regularly presented by Egret in the "Simbad News" papers in the CDS Information Bulletin. We will just recall here, for completeness, the main features of this astronomical object-oriented database:

 ▷ a database of more than 1 million astronomical objects (stars, galaxies and all astronomical objects outside the solar system);

▷ a cross-index to several hundreds catalogues including observation logs of space missions such as IRAS (see Preite-Martinez, 1993, for cross-identification considerations) or ground-based catalogues, such as the recently integrated Positions and Proper Motions Catalog (PPM);

▷ pointers to catalogued data (some 25 different types of data measurements) and bibliographic references covering the complete astronomical literature since 1950 for stars, and since 1983 for extragalactic objects;

▷ an interactive object-oriented database system offering several user interfaces (command-line, Xsimbad, ADS service, e-mail batch mode, *etc.*), all of them using a common client/server mode for remote access;

▷ an efficient management of the possible variations in the naming of astronomical objects with the *sesame* module and the *info* database of nomenclature of celestial objects (see section 6.2 below);

▷ a name resolver integrated within other applications (ESIS, STARCAT, STARVIEW, ISSA-PS, HEASARC, IRSKY, *etc.*), taking benefit of the client-server approach previously mentioned.

SIMBAD is a charged service. Users need to register, and get a userid/password from the CDS staff (or from the U.S. agent for American users). The charges are covered by NASA for all U.S. users, and, starting January 1995, are covered by ESO and ESA for all European users from ESO or ESA member states.

SIMBAD is a database evolving day after day:

▷ New data (bibliographical references, identifiers, basic data), and new acronyms are being entered on a daily basis; this is done as a result of a continuous survey of the astronomical literature, under the responsibility of Institut d'Astrophysique de Paris with the collaboration of Paris and Bordeaux observatories (see section 5.1 below).

On the other hand, large astronomical catalogues with their own identifiers and measurements are added after a cross-matching procedure which frequently spans over many months (examples of catalogues recently integrated or still under scrutiny are: PPM , Hipparcos Input Catalogue, IUE Log of observations, CCDM , *etc.*).

In 1994 about 500 new bibliographical references per month were added to SIMBAD, while the rate of growth of the number of objects in SIMBAD was about 2500 new objects monthly.

▷ The data contained in SIMBAD are also permanently updated, as a result of errata, remarks from the librarians (during the scanning of the literature), quality controls, or special efforts from the CDS team to better cover some specific domains (e.g., in 1994, multiwavelength emitters and complex objects). Requests for corrections, errata, or suggestions are regularly received from SIMBAD users through a dedicated e-mail address (`question@simbad.u-strasbg.fr`). In 1994, about 10 mails per week concerning updates of SIMBAD were received at the CDS. Corrections of errors are made under the responsability of CDS astronomers coordinated by Gérard Jasniewicz.

▷ Working groups at the CDS prepare continuous improvements of the SIMBAD database. Three working groups have elaborated new concepts which will eventually become available in 1995:

  ◇ Hierarchy and links between objects in SIMBAD. The hierarchy will be used for displaying components of complex objects (multiple stars, star clusters, etc.). Links become necessary for pointing out possible relationships between objects observed at various wavelengths.

  ◇ New basic data. The revision of the current set of basic data comes from the necessity to provide the user with the source of data and to add data collected in various wavelength ranges.

  ◇ New object types. Several object types appear to be frequently necessary in order to categorize a given astronomical object .

Together with the building of the SIMBAD database, the CDS expertise on cross-identification of astronomical sources, and more generally on scientific data handling, allow helpful contributions to large space and ground-based projects. To name a few projects in which the CDS has played (or plays) some role: IRAS, IUE (see *e.g.* Egret *et al.* 1992b), HIPPARCOS and TYCHO (see *e.g.* Turon *et al.* 1991, Egret *et al.* 1992a), ISO, DENIS, *etc.*

16–4   The ALADIN interactive sky atlas

Astronomers looking for optical counterparts of gamma-ray to radio sources, need to go beyond already existing catalogues or databases. Major international efforts for providing digitizations of sky surveys (see *e.g.* , McGillivray, 1994), together with recent technological developments in the data storage capacity (see chapter in this volume, page 243), now allow to plan a complete on-line digitized sky atlas.

ALADIN is a new project currently under development (Paillou *et al.* , 1994a, 1994b), to create an interactive atlas of the digitized sky allowing the user to visualize on his/her own workstation digitized images of any part of the sky, to superimpose entries from astronomical catalogues or user data files, and to interactively access the related data and information from the SIMBAD database for all known objects in the field.

The software architecture of ALADIN is based on the client/server philosophy (Bonnarel *et al.*, 1994). Each set of stored data (astronomical catalogues, SIMBAD database, and image pixels) are accessed through a dedicated server.

This new tool will be particularly useful for multi-spectral approaches (*e.g.*, searching for counterparts of sources detected at various wavelengths), and for a number of applications related to the database quality control and the cross-identification of observational data.

Currently (October 1994), ALADIN is only available as a local prototype; it will eventually become a public interactive tool, available for all laboratories through networks. The image database, when completed, will include the digitized sky survey produced at STScI, together with full resolution digitized plates in regions of confusion (about 500 Gbytes on-line).

16–5 Bibliography and literature search

The bibliography of objects is one of the unique features of the SIMBAD database. This service is gradually being extended in order to provide the user with a wider perspective of the current astronomical literature.

5.1 THE SIMBAD BIBLIOGRAPHY

The SIMBAD bibliography for the astronomical objects includes references to all published papers from the largest coverage of astronomical journals (currently about 85 titles; see Table 2 of the chapter 24 in this book). Articles are scanned in their entirety, and references to all objects mentioned are included in the bibliography. References, authors, and titles are stored for more than 80,000 papers since 1950 and some 1,4 million references to astronomical objects.

The updating of SIMBAD from the published literature is a continuous daily process performed through a collaboration with the Institut d'Astrophysique de Paris and the Paris and Bordeaux observatories (Laloë *et al.*, 1993).

5.2 ABSTRACTS OF RECENT PAPERS

More recently, through agreements with the Editors, the abstracts of Astronomy & Astrophysics, main journal and Supplement Series, as well as the abstracts from the Publications of the Astronomical Society of the Pacific (PASP), are made available on-line a few weeks before publication.

The collection of abstracts from these two journals, starting from January 1994, is available on the World-Wide Web, and can be queried by keywords or author names. Links are made between the abstracts and the electronic tables, when available. The use of a common bibliographic 19-digit reference code (see chapter in this volume, page 259) also shared with NED and ADS, allows easy cross-indexing between various systems.

This bibliography service is expected to be developed in the future, with the planned inclusion of other abstract collections, and the implementation of advanced search mechanisms.

The analysis of bibliography linked to astronomical objects is an efficient tool for scientific investigations and database quality control (as shown by Lesteven, 1994). New methods to be developed in complement to the bibliography service will eventually include the use of neural networks for document classification and information retrieval, and the automatic generation of hypertext. These tools will be especially useful in the perspective of full text searches, with the forthcoming development of electronic publication and, more generally, electronic information handling (see, *e.g.* , Heck, 1992, 1995b).

16–6   Yellow–Page Services

The number of on-line astronomical services (databases, datasets, catalogues, archives, ftp-services, gophers, WAIS indexes, information systems, directories of services, Web home pages, etc.) is such that the potential user critically needs directories and even meta-directories, in order to discover the resource where answers to his/her scientific requests can eventually be found.

In this context, the effort to combine yellow-page services and meta-databases of active pointers is a crucial solution to the data retrieval problem.

The concept of yellow-page services was probably publicly introduced for the first time in the astronomical community at the ALD-II Conference in Haguenau (Heck & Murtagh, 1992).

6.1   THE STAR*S FAMILY

This is the generic name for a growing collection of directories, dictionaries and databases which is described in more details in a separate chapter in this volume (see page 195). The following databases are accessible via the World-Wide Web CDS server [→1]:

StarWorlds: Addresses, telecom plugs (including active URLs for WWW access) and all other practical data of about 5000 organisations of interest to astronomers and related space scientists.

StarBits: Dictionary of about 80,000 abbreviations and acronyms.

StarHeads: A database of about 1000 personal web pages of astronomers and related space scientists.

6.2 ASTROWEB

AstroWeb (Jackson *et al.* 1994) is a collection of pointers to astronomically relevant information resources available on the Internet. It is maintained by the AstroWeb consortium, a group of scientists from CDS (Strasbourg astronomical Data Center), MSSSO (Mount Stromlo and Siding Spring Observatories), NRAO (National Radio Astronomy Observatory), STScI (Space Telescope Science Institute) and ESO/ST-ECF (European Southern Observatory, Space Telescope–European Coordinating Facility).

The AstroWeb database provide hypertext links (anchors) to more than 1200 Internet resources (January 1995).

16–7 Conclusion: Towards a global astronomical data and information service

The various activities of the CDS, as described above, are all directed towards the organization of simple and unified views of astronomical data and information. The trend in the coming years will be towards interfacing CDS services (for instance SIMBAD and the catalogue server within ALADIN) in order to provide even more powerful tools to the users (Figure 16–1).

These new developments imply a dedicated effort in terms of research, in order to derive the best solutions according to the existing information technology.

Examples of such efforts are:

▷ the research on the concept of a *Reference Directory* for the ESIS project (Egret *et al.* 1990);

▷ the research on new methods of information handling and quality control (S-space, neural networks, hypertext approach);

▷ prototype work on the World-Wide Web service;

▷ the organization of workshops and conferences directed towards the astronomical community, such as "Weaving the Astronomical Web", a Conference held in Strasbourg, 6-7 April 1995.

THE CDS WWW SERVER

A CDS information service on the World-Wide Web [→1] has been publicly available since December 1993. It provides the following pages:

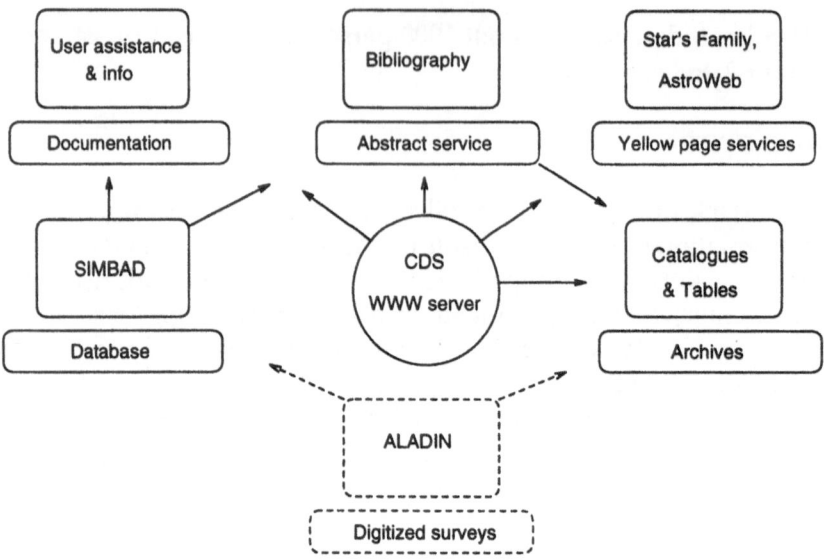

Figure 16–1: Pictorial user's view of CDS services and their underlying information holdings (arrows show existing client/server links.)

- ▷ General information on CDS
- ▷ Access to the Catalogue service
- ▷ SIMBAD documentation
- ▷ Abstract service
- ▷ On-line version of the CDS Information Bulletin
- ▷ Yellow-Page services including the Star*s Family and AstroWeb
- ▷ Access to the TOPBase of the Opacity project
- ▷ *etc.*

More details about CDS activities can also be found in the recent issues of the six-monthly CDS Information Bulletin.

Contact Information

Send all requests, remarks, or comments to the electronic mail address:

`question@simbad.u-strabg.fr`

References

[1] Bonnarel, F., Divetain, E., Ochsenbein, F., Paillou, Ph., Wenger, M. (1994), in *Astronomy from Wide-Field Imaging*, Postdam, Germany, H.T. MacGillivray Ed., Kluwer Acad. Publ., 353.

[2] Crézé, M. (1993), *Vistas in Astronomy* **36**, 163.

[3] Egret, D., Ansari, S.G., Denizman, L., Preite-Martinez, A. (1990), "The ESIS Query Environment", ESIS Internal Report, CDS, Strasbourg, 53 pp.

[4] Egret, D., Wenger, M., Dubois, P. (1991), in *Databases & On–line Data in Astronomy*, Albrecht & Egret (Eds.), Kluwer Acad. Publ., 79.

[5] Egret, D., Didelon, P., McLean, B.J., Russel, J.L., Turon, C. (1992a), *Astron. Astrophys.* **258**, 217.

[6] Egret, D., Jasniewicz, G., Barylak, M., Wamsteker, W. (1992b), in *Astronomy from Large Databases II*, Proc. Coll. Strasbourg - Haguenau, Eds. A. Heck & F. Murtagh, ESO Conf. & Workshop Proc. **43**, 265.

[7] Egret, D., Ochsenbein, F. (1994), CDS Inform. Bull. **44**, 57.

[8] Grosbøl, P. (1991), in *Databases and On-line Data in Astronomy*, M.A. Albrecht & D. Egret (Eds.), Kluwer Acad. Publ., 253.

[9] Heck, A. (1992), "Desktop Publishing in Astronomy and Space Sciences", Strasbourg, World Scientific, Singapore, xii+240 pp.

[10] Heck, A. (1995a), "The Star*s Family – An example of comprehensive yellow-page services", in *Information & On-line Data in Astronomy*, D. Egret and M. A. Albrecht, Eds., this book, page 195.

[11] Heck, A. (1995b), "Facets and Challenges of the Information Technology Evolution", in *Information & On-line Data in Astronomy*, D. Egret and M. A. Albrecht, Eds., this book, page 1.

[12] Heck, A. (1995c), *Astron. Astrophys. Suppl.* **109**, 265.

[13] Heck, A. and Murtagh, F., Eds. (1992), *Astronomy from Large Databases II*, Proc. Coll. Strasbourg - Haguenau, ESO Conf. & Workshop Proc. **43**.

[14] Heck, A., Egret, D., Ochsenbein, F. (1994), *Astron. Astrophys. Suppl.* **108**, 447.

[15] Jackson, R., Wells, D., Adorf, H.M., Egret, D., Heck, A., Koekemoer, A., Murtagh, F. (1994), *Astron. Astrophys. Suppl.* **108**, 235.

[16] Laloë, S., Beyncix, A., Borde, S., Chagnard-Carpuat, C., Dubois, P., Dulou, M.R., Ochsenbein, F., Ralite, N., Wagner, M.J. (1993), *CDS Inform. Bull.* **43**, 57.

[17] Lesteven, S. (1994), "Méthodes d'analyse multidimensionnelle appliquées à la recherche d'information bibliographique: Contrôle-Qualité d'une base de données astronomiques", Thesis, Université Louis Pasteur.

[18] Lortet, M.C., Borde, S., and Ochsenbein, F. (1994), *Astron. Astrophys. Suppl.* **107**, 193.

[19] MacGillivray, H.T., Editor (1994), "Astronomy from Wide-Field Imaging", Postdam, Germany, Kluwer Acad. Publ.

[20] Ochsenbein, F. (1994), *CDS Inform. Bull.* **44**, 19.

[21] Paillou, Ph., Bonnarel, F., Ochsenbein, F., Crézé, M. (1993), "Aladin Deep Sky Mapping Facility", Project Report, CDS, Strasbourg, May 1993.

[22] Paillou, Ph., Bonnarel, F., Ochsenbein, F., Crézé, M. (1994), in *Astronomy from Wide-Field Imaging*, Postdam, Germany, H.T. MacGillivray Ed., Kluwer Acad.

Publ., 347.

[23] Paillou, Ph., Bonnarel, F., Ochsenbein, F., Crézé, M., Egret, D. (1994), in ADASS III, ASP Series *61*, 215.

[24] Preite-Martinez, A. (1993), *Bull. Inform. CDS* **42**, 61.

[25] Turon, C., Arenou, F., Baylac, M.-O., Boumghar, D., Crifo, F., Gómez, A., Marouard, M., Morin, D., Sellier, A., (1991) in *Databases & On–line Data in Astronomy*, Albrecht & Egret (Eds.), Kluwer Acad. Publ., 67.

[26] Wells, D. (1992), in *Astronomy from Large Databases, II*, ESO Proceedings **43**, 143.

[27] Wells, D. C., Greisen, E. W., Harten, R. H. (1981), *Astron. Astrophys. Suppl.* **44**, 363.

Access Pointers

[→1] CDS: `http://cdsweb.u-strasbg.fr/CDS.html`

17

Data Holdings in the ADS

Stephen S. Murray, Guenther Eichhorn, Alberto Accomazzi,
Carolyn Stern Grant & Michael J. Kurtz

17-1 Introduction

The ADS has been restructured recently (October 1994) so that it will emphasize the very successful Abstract Service and evolve to provide more Digital
Library capability. It was also decided to make World-Wide Web HTTP protocols the primary means of implementing the ADS services. A first step in
this direction has been the development of the Astrophysics Science Information and Abstract Service (ASIAS) as a Web service [→1]. We have added
to ASIAS an Article Service that provides access to scanned images of Astrophysical Journal Letters. These literature data holdings include over 160,000
astronomy related abstracts as generated by the NASA RECON office and
ten years of Astrophysical Journal Letters from 1984 through 1994. We expect to complete scanning of the Letters back through 1975 shortly. We will
also scan new columns as they are published. A lag time of a few months
(maximum) is expected. We are working with the AAS and *Astrophysical
Journal* so that we will be able to provide access to the full Journal, and we
plan to also scan other astronomy and astrophysics journals as agreements
are reached with their publishers.

17-2 The Astrophysics Science Information and Abstract Ser-
vice (ASIAS)

The ASIAS is accessed through a forms interface. The ASIAS Query Form
[→1] provides the user with easy access to the capabilities of the abstract
service. The top part of the form provides spaces to specify the query parameters, like author names, publication date limits, titles, etc. The bottom
part of the form allows the user to configure the abstract service search engine. For instance it allows to select whether a match in a particular field is
required or optional for selection of the final list. This allows the user for

D. Egret and M. A. Albrecht (eds.), Information & On-Line Data in Astronomy, 175–184.
© 1995 *Kluwer Academic Publishers.*

instance to query only for papers written by a certain author. A complete description of these capabilities is given in the Abstract Service on-line Help.

2.1 ABSTRACT QUERIES

The abstract service returns a list of abstracts. This list includes the bibliographic codes, the score of the abstract in the search, the publication date, the list of authors and the title. For instance a query with "Eichhorn, G" in the author field and "gamma ray detectors in antarctica, SN 1987A" in the abstract text field with the default settings returns a list of abstracts with different scores. Two abstracts have a score of 1.0 since all the words were found in an abstract with the specified author. Subsequent references have progressively smaller scores, depending on how many words were matched.

The bibliographic codes in this list are links to the full abstracts. It is also possible to retrieve several abstracts at once by checking the checkbox next to the selected references and clicking on the "Retrieve selected abstracts" button at the bottom of the page. This retrieval allows the user to select either screen viewing, sending to a printer or saving to a local file.

For the publications of the AAS (Astrophysical Journal, Astrophysical Journal Letters, and Astronomical Journal) we provide the capability to get a table of contents. This form allows access to a list of publications by year, month, or volume number.

2.2 LIST QUERIES

This form allows the user to retrieve lists of data items in the abstract service system like lists of authors>, keywords, or synonyms. It also allows queries to the SIMBAD database at the Centre de Données astronomiques de Strasbourg (CDS) to find synonyms for astronomical object names. The lists are accessible through hyperlinks over the text entry boxes on the query form or through a separate list query form.

2.3 FULL ARTICLE IMAGES

For some of the papers we have the images of the full articles on-line. If an article is available for a selected abstract, links to the images are available in the returned abstract. An example of this can be found when retrieving the second abstract in the previous list. These links allow the user to either view the pages on the screen or to send the images with different resolution directly to the printer.

The screen view images are in GIF format with resolution of 75 dots per inch (dpi) in 4 level greyscale. This makes them readable on the screen. However, this resolution is not good enough for printing. For printing there are two versions available: a version with a resolution of 150 dpi and one with 300 dpi. The 150 dpi version prints very fast, but is not quite as readable, especially for plots and figures. The higher resolution version prints slower, but improves the image quality.

The higher resolution versions are available as postscript level 1 or level 2 files and as TIFF files with G4 compression. The link to "More Article Retrieval Options" allows to retrieve different versions and either print them, view them, or save them on the local computer.

These images are also accessible directly through the Reference Query Page if the user knows the journal reference of a paper. Here, the user specifies the journal, volume and page of an article to retrieve the bitmapped images.

2.4 DIRECT ACCESS VIA BIBLIOGRAPHIC CODES

This access method allows access to the ASIAS services directly through bibliographic codes or NASA/STI accession number. The bibliographic code can be constructed if the reference of an abstract is known and allows direct access to this abstract as long as it is in the database.

Both the abstracts and the full article images are available directly for instance for building links in other documents. The URL to use for the direct access to abstracts is:

```
http://adsabs.harvard.edu/cgi-bin/bib_query?"bibcode"
```
and for full article images it is:
```
http://adsabs.harvard.edu/cgi-bin/article_query?"bibcode"
```
where "bibcode" is the bibliographic code for the requested abstract or article, or its NASA/STI accession number. This allows any data provider to link to our database and include our data in other documents. This can be used for instance to build reference lists in electronically published articles that contain links to the referenced papers. A bibcode verification utility is available to check bibcodes for correctness and availability of abstracts and full articles.

2.5 QUERY FEEDBACK

The query feedback mechanism allows the user to automatically formulate a new query from the retrieved abstract. This is an easy way to do exhaustive literature searches for a given subject. The user specifies which parts of the current abstract should be used for a new query and then executes this query without having to transfer data from one window to another. The query feedback form is linked to the bottom of every abstract that is retrieved.

As an example you can look at the abstract from the previous example about Gamma Ray Observations from Antarctica. This abstract has links to the full article as well as the query feedback form at the bottom. The query feedback form allows to either execute the feedback query immediately or to return a new form that has the specified fields filled with the information from the current abstract. This allows further modifications to the feedback query before sending it off.

2.6   DATA ACCESS

The Centre de Donnés astronomiques de Strasbourg (CDS) provides access to data tables published in some astronomical articles. We have interfaced with the CDS system to provide links from our abstracts to their data tables whenever they are available for a selected abstract. An example can be found with a bibcode query for 1982ApJS...49...27U, an article in the 1982 *ApJ Supplements*. The full abstract has a link to the data tables, which can be retrieved directly from the CDS.

2.7   CONTRIBUTIONS FROM OTHER SOURCES

We are currently rebuilding our database system for the abstracts in order to be able to include abstracts from other sources. We would welcome any contributions of abstracts from journals that are not or only incompletely covered, as long as they are within our subject area and conform to our data format. If you have abstracts that you want to contribute, please contact the ADS project at ads@cfa.harvard.edu.

17–3   Astronomical Catalogs on the ADS

In addition to the literature, the ADS Project has been developing WWW access mechanisms to many of the astronomical data catalogs and NASA Mission data sets that were part of the "Classical ADS". We have developed a form–based WWW service that will search any of more than 150 catalogs and return results from queries to users in a variety of formats. The list of catalogs presently available can be found on-line at [→2]. These are not necessarily resident on systems operated or maintained by the ADS, but are links to catalogs that are being made accessible via ADS WWW servers that are operating at remote sites.

For the catalogs listed in [→2] the ADS provides documentation that describes the catalog, the contents and format of the data, and sample queries. The ADS has developed guidelines and templates for documenting astronomical catalogs that are helpful to both data providers and users. Adding new catalogs to this service is made easy though the use of these documentation templates. There is also a WWW SQLServer package that the project offers to data providers. This service dynamically constructs an HTML query form for a catalog when the user decides to initiate a catalog search. The form is based on a catalog description file that is maintained at the catalog site. Thus, the maintainers of the catalogs are free to make changes to their data as needed, and as long as the descriptions and descriptor file are up to date, the ADS will automatically remain current.

When the query form is filled out, and the query issued, ADS provided software (a CGI script running on the ADS WWW server) translates the request into an SQL statement that is passed on to the site where the catalog database resides. The SQLServer, customized and installed at the site, parses the SQL statement into whatever commands are required to perform the search. In many cases this transformation is merely a pass-through of the SQL statement. Customization usually consists of identifying the name of the local DBMS process. If SQL is not supported by the database management system, then there will also be a module in the SQLServer to generate appropriate commands. Results are temporarily held at the data catalog site and are reformatted by the SQLServer into the selected format requested by the user. The supported formats are listed in Table 17–1. Note that the PostScript finding chart format is only applicable to the Minnesota Palomar Observatory Sky Survey (APS_POSS) catalogs. The files returned can easily be imported into standard astronomical data analysis programs for further use.

ASCII Text	as an HTML page or as a file
FITS ASCII Table	as an HTML page or as a file
ADS Table	as an HTML page or a file
Postscript	finding charts for APS_POSS

Table 17–1: ADS Catalog Query Result Formats

17–4   Using the ADS Catalog Service

Access to the ADS Catalog Service is via the WWW at [→2]:

```
http://adswww.harvard.edu/ads_catalogs.html
```

Users can obtain lists of the catalogs in the ADS Service and then use the query forms to make selections from these catalogs. In most cases it is possible to make selections based on any field in the catalog and to format the results as desired for further use. In the Classical ADS the client software installed at the user workstation allows multiple queries to be sent to several databases and provides a workspace for integrating the results and such services as coordinate conversion, plotting, and correlating. In the WWW version there is no ADS client process running on the user workstation, and there is no workspace maintained by the Web Browser. Development work with NCSA on the Common Client Interface (CCI) is aimed at addressing some of these issues. We are especially interested in coordinate conversion "on the fly" so that correlations of different catalogs are easier.

The descriptive information collected about catalogs is illustrated in Table 17–2. Our experience has been that users not familiar with a particular catalog need such basic information as the coordinate system that is used for object positions, the units that are used for source properties, and examples of catalog queries. The templates we have developed give data providers an easy mechanism for generating these descriptions, and they also give users a consistent "look and feel" to these data.

Description	A description of this database
Field List	A general listing of all the fields in the database
Example	Example SQL query to the database with results
Filename	The URL of this document
Origin	Origin of the data in the database
Date	Creation date
Revised	Last revision date
Columns	Number of columns or fields in the database
Rows	Number of rows or entries in the database
Time Period	Time period for which catalog data are available
Other	Some additional information
Contact	Addresses for further information

Table 17–2: Catalog Information

The description for a catalog is brief and helps users determine if the information is relevant to their needs. For example, the EUVE bright source catalog description (Christian *et al.*) is as follows:

> The catalog that corresponds to this documentation contains a detailed list of verified bright sources detected during the survey phase of the Extreme Ultraviolet Explorer (EUVE) mission (calibration targets are also included). Two distinct surveys, the All-Sky and Deep Surveys, were conducted by the four EUVE telescopes during the first six months of the mission, as well as during short periods of "gap filling" during the first six months of the Guest Observer (GO) program. A companion database, euve_cat1supp, contains a

list of sources that appeared in the Bright Source List (BSL) but that were not formally included in FESC because they did not meet the criteria for inclusion.

The All-Sky Survey was conducted by three scanning telescopes ("scanners") with four distinct filters (Lex/B, Al/C, Ti/Sb/Al or "Dagwood", and Sn/SiO) covering the wavelength region 58-740. During the All-Sky Survey phase, the scanning telescopes mapped approximately 97% of the entire sky with exposure times of greater than 20 seconds along the ecliptic and much higher at the poles. Concurrently, the Deep Survey (DS) telescope conducted a much more sensitive survey of a 2 Degrees x 180 Degrees band along the ecliptic in two filters (Lex/B and Al/C) covering the wavelength region 67-364. The catalog itself, as well as descriptions of the data analysis procedures, will also appear in the "The First EUVE Source Catalog" by S. Bowyer, et al. (1994), which is published in the August edition of Astrophysical Journal Supplement. For more information concerning calibration targets, "The Extreme Ultraviolet Explorer Bright Source List" by R. Malina, et al. (1993), published in the Astronomical Journal.

For a complete description of the telescope instrumentation on the science payload see the "EUVE Guest Observer Program Handbook" (NASA NRA #93-OSSA-X, Appendix G, Chapters 1 and 2).

The field descriptors are one or two line explanations of the field contents, and include formats and examples. Table 17–3 shows examples taken from the EUVE entry.

Field	Description	Example
euve_name	EUVE name of source	char20:"EUVE_J0007+33."
ra (deg)	Right Ascension of EUVE source	real (epoch 2000): 1.8833
dec (deg)	Declination of EUVE source	real (epoch 2000): 33.2983
lex (counts/s)	All-Sky Lexan/B filter countrates	real: 0.219
lexerr	All-Sky Lexan/B filter statistical uncertainty	real: 0.014
alc (counts/s)	All-Sky Al/Ti/C filter countrates	real: 0.074
alcerr	All-Sky Al/Ti/C filter statistical uncertainty	real: 0.009
dag (counts/s)	All-Sky Ti/Sb/Al filter countrates	real: 0.221
dagerr	All-Sky Ti/Sb/Al filter statistical uncertainty	real: 0.022

Table 17–3: Field Descriptor Examples from EUVE

For each field, there is a more detailed description that is hyperlinked to the field name. Examples are shown in Table 17–4.

Column	Description
euve_name	Name of the EUVE source. The adopted format is EUVE_Jhhmm[+-]dd.d which designates an EUVE source located at the approximate position of hhmm[+-]dd.d (hours and minutes of Right Ascension and decimal degrees of Declination, respectively.) This position is in Epoch 2000 ("J") coordinates and is based on internal calculations. Note that the format used in this database is slightly different from that used in the published version of the catalog, where the declination in the EUVE name is multiplied by ten. Example: (euve_name) (char20): "EUVE_J0007+33." Example: (euve_name) (char20): "EUVE_J2353-70.3"
ra dec	The Right Ascension and Declination in decimal degrees of the EUV source (Epoch 2000). Position is accurate to within one arcminute. Example: (ra) (real): 349.882996 Example: (dec) (real): 79.011703

Table 17–4: Column Descriptions

From the Catalog Service HomePage, the ADS also provides links to other catalog services that are not directly supported by the project. These include those listed in Table 17–5.

The ADS will add links to other catalogs as requested. However, we encourage catalog providers to work with the ADS to include access to their holding using the ADS Query Forms interface and WWW SQLServer. This access may be in addition to direct access through the site.

17–5   Data Archives

The WWW version of ADS provides a forms based access to data archives as well as catalogs. The model used is taken from the Classic ADS. Archive queries are a two step process. An archive metadata database is queried to produce a list of possible archive files for transfer. This query is essentially identical to a catalog query, except that the returned list is an HTML document that includes hyperlinks to the archival data files. Selecting such a link results in a file transfer to the user. At this time there are two archives available via the ADS WWW Archive Service, they are listed in Table 17–6.

As with catalogs, the ADS WWW Archive Service HomePage provides links to other archive services on the Web. Table 17–7 lists those currently

Source	Description
Univ. of Massachusetts	The new Reipurth General Catalog of Herbig-Haro Objects
	Herbig-Bell Third Catalog of Emission-Line Stars of the Orion Population
	The D'Antona and Mazzitelli Pre-Main Sequence evolution tracks
	The Kleinmann-Hall 2 micron spectra of MK standards and a few hundred spectra of stars in Praesepe and M67
	The entire archive of the Star Formation Newsletter ed. B. Reipurth
HEASARC	HEASARC Catalogs
IUE	IUE Merged Log of Observations
NIST	Atlas of the Spectrum of a Platinum/Neon Hollow-Cathode Reference Lamp in the Region 1130-4330 A

Table 17–5: Links to Other Catalogs

Source	Description
SAO	Einstein Archive Service
NIST	Atomic Spectroscopic Database (Beta Version)

Table 17–6: ADS WWW Archive Service

included. The ADS will add links as requested. We are also interested in adding more archives to the Service and will work with data providers to help create searchable metadata databases and appropriate query forms.

Source	Description
IPAC	The ISSA Postage Stamp Server
HEASARC	StarTrax: The HEASARC WWW Archive Interface
NSSDC	Skyview: A Virtual Observatory

Table 17–7: Links to Other Archives

Access Pointers

[→1] ASIAS (Abstract service):
    `http://adsabs.harvard.edu/abstract_service.html`
[→2] ADS Catalog service: `http://adswww.harvard.edu/ads_catalogs.html`

18

The European Space Information System (ESIS)

Paolo Giommi & Salim G. Ansari

18–1 Introduction

ESIS was conceived in the late 1980's as a service to provide a uniform view over intrinsically different data services and to support the comparison of data coming from different archives.

The recent advent of the World Wide Web significantly simplified on-line access to astronomy data. However some basic answers such as uniformity of access and data comparison remain largely unanswered.

The ESIS system is now operational and provides the planned broad-band, uniform view of astronomy data from several electronic archives. It also provides tools to compare data from more than one experiment or database. Most of the ESIS services are available both from dedicated client software and from applications running on the World Wide Web. Detailed analysis cannot be performed within ESIS. It must be done in specialized environments depending on the type of data considered. Some of these environments can be started directly from ESIS.

In the following sections we describe in some detail the concepts, the architecture, and the various services available within the ESIS system as of late 1994.

18–2 ESIS from the World-Wide Web

The World-Wide Web (WWW), a project that began at CERN, provides a simple client-server system which uses a method of textual hyperlinks to navigate through *networked pages* of information. Because it is simple to set up a WWW server, and because WWW utilizes the standard TCP/IP networking protocol, this tool is very attractive to archive sites wishing to

D. Egret and M. A. Albrecht (eds.), Information & On-Line Data in Astronomy, 185–193.
© 1995 *Kluwer Academic Publishers.*

make their services available to the community *without the need to build their own client-server applications on multiple platforms.* The advantage lies more in the uniformity of access to a variety of services (FTP, Gopher, WAIS, Archie, *etc.*) rather than the need to download and install a client for each service.

In November 1993, ESIS began to provide a homepage on the World-Wide Web, which gave details on how to use the ESIS system, including information on the project, newsletters, and a user's guide to ESIS. With the advent of NCSA Mosaic Version 2.0 in December 1993, which implemented the first form-based interface, it was possible to go much further. The ESISBIB service was put on-line soon after. In February 1994 ESIS began to provide a catalogue browser to the more than 60 catalogues and mission logs (Ansari *et al.* , 1994a).

The World-Wide Web is very general in its concept. It allows users to navigate through massive amounts of information and data. Because it is not aimed towards a single community, it cannot fulfill the needs of everybody. Therefore, the next logical step was to review some of the applications that were already being used by the community. Only this time ESIS began to take advantage of the hypertext transfer protocol (http) on which the World-Wide Web is based. If we assume that NCSA Mosaic is a Graphical User Interface of the World-Wide Web, we can also assume that the conformity to the `http` protocol in other community-specific applications can become so general in the access of data that building such applications can concentrate more on the value-added scientific functionality rather than concentrate on the network protocol architecture. The ESIS project recently began investigating the feasibility of developing its applications around the hypertext transfer protocol. And the activity showed immediate success. The only drawback of this access method is that the hypertext transfer protocol is *stateless* —each search triggers a single action on the server side, which executes a single command and sends the answer back to the client. In order to refine or continue the search for data other methods must be implemented to preserve the state of the search.

All ESIS components today depend on the `http` protocol. In the following section we describe each and every component and its functionalities. The ESIS Homepage and some of these services is available from the Uniform Resource Locator (URL) [→1].

18–3   The ESIS System

The present version of the ESIS software is composed of a complex core of applications that give a uniform view over intrinsically different databases. A wide variety of astronomical catalogues and data products (*i.e.* images, spectra, etc.) are available from several databases. Currently data products can be retrieved from the following missions: EXOSAT, HST (preview data, from DAO, Canada), IUE, *Einstein*, ROSAT, IRAS, and COS-B.

The following general services are available

▷ The Catalogue Browser

▷ The Imaging Application

▷ The Spectral Application

▷ The Timing Application

▷ The Bibliographic Service

All these applications are connected via computer networks to several archive centers from which the data can be retrieved on-line or requested for off-line delivery. The ESIS applications makes all the connected databases look like a single large distributed archive.

3.1 THE CATALOGUE BROWSER

The Catalogue Browser is that part of ESIS where astronomical catalogues, data product lists, and mission logs can be browsed and compared in detail. There are several dozen catalogues available. One or more catalogues can be searched at a time. Catalogues can be selected one by one or by class, wavelength region, mission name or using keywords (*e.g.* , keyword "redshift" will select all catalogues including redshifts of extragalactic objects). There are five main groups of catalogues: astronomical catalogues; data product listings; mission logs; result databases; and target lists.

Three types of searches can be performed:

SEARCH IN A CONE. A search around a specified position in the sky and within a given radius is performed. If the user supplies a name of a cosmic source (instead of its celestial coordinates) the system automatically retrieves its coordinates from the SIMBAD database.

SEARCH BY NAME. All catalogue entries corresponding to a given name, (or with any of its *aliases* as obtained from SIMBAD) are retrieved.

SEARCH BY PARAMETER. Conditions on any parameter of one or more catalogues can be specified using a simple syntax. For example the condition RA > 20 AND DEC < -50 AND VMAG > 15 will retrieve all the entries with right ascension greater than 20 degrees and declination less than -50 degrees and with visual magnitude greater than 15.

A special kind of search is the *overview* where all the entries concerning the observations of a given object are retrieved and a graphical interactive window is built giving a visual general overview of the data in the frequency-observation date plane. The user can click on the various buttons to retrieve more details about the observations or the actual data without having to know where the data are located or the format used.

Catalogues can also be *joined* (cross-correlated) simply by clicking on the selected tables on a form and specifying the radius to be used.

Most of the search capabilities of the ESIS catalogue browser are also available within the data visualization and manipulation applications (i.e. imaging, timing and spectral applications). The only difference is that the catalogues that can be accessed are only those which are relevant for the application. For instance from the imaging application only catalogues concerning astronomical images are available. Once the query has been completed, clicking on a retrieved entry will automatically copy the corresponding file from the remote archive and load it into the visualization package.

3.2   THE IMAGING APPLICATION

The imaging application is that part of ESIS where the user can retrieve, display and manipulate imaging data from the EXOSAT, *Einstein*, ROSAT, HST and COS-B databases. Many other types of astronomical images like optical or radio FITS images, can also be read as local files, displayed and compared with the data available within the system. Detailed analysis is not supported, however if this is necessary the user can easily escape from the ESIS application into more specialized analysis software.

The main features of this applications are:

▷ retrieve and display images,

▷ display images from different experiments on the same scale,

▷ overlay source positions from one or more catalogue(s) available within ESIS and other systems,

▷ overlay iso-intensity contours,

▷ generate image mosaics.

A summary of all the images available can be requested simply by giving a source name. The system automatically issues a query to the SIMBAD database which returns the source coordinates to be used to extract the list of all the images available.

The imaging application is made up of three main components:

1. a Graphical User Interface;

2. a command line image manipulation program called XIMAGE (Giommi *et al.* 1992). XIMAGE was originally developed for the EXOSAT database system and has been upgraded to support other data types as a collab-

orative effort between ESIS and the High Energy Astrophysics Science Archive Center (HEASARC);

3. a set of http-specific functions to retrieve catalogue entries from the main ESIS server by use of the World-Wide Web client/server protocol.

The GUI receives commands from the users and translates them either into the XIMAGE native language or connects to the ESIS server to retrieve catalogue entries. XIMAGE then executes the commands and displays the results in the graphic display.

3.3 THE SPECTRAL APPLICATION

The ESIS spectral visualization and manipulation package provides access to HST GHRS and FOS spectra (made available by the Canadian Astronomical Data Centre, CADC), the IUE low–dispersion spectra of the Uniform Low Dispersion Archive (ULDA, Wamsteker *et al.* 1989) and EXOSAT ME and GSPC X-ray spectra.

In addition ROSAT and *Einstein* X-ray spectra can be extracted from event files from the imaging application and displayed within the spectral package.

As in other ESIS packages, the user may select data by searching and identifying spectra from catalogues that contain spectral product information. The package would then fetch the selected spectrum from a remote archive and display it to the user.

One of the features of the spectral package is the possibility of building a multi-frequency (radio-to-X-ray) energy distribution for any given object. ESIS retrieves all the individual flux values from its databases and plots the energy distribution after converting all fluxes to common units.

Other features include the parabolic and Gaussian fitting to spectral line profiles, which are used to determine the central wavelength. The system consults the Kurucz & Peytremann (1975) wavelength list, which then displays the chemical elements, their wavelength, oscillator strengths and excitation potential.

The graphical user interface, based on the Athena Widget Plotter (Klinge-biel, 1992), can display spectra on a uniform scale (which is automatically rescaled whenever a new spectrum is read in). This particularly useful feature allows the user to display and cross-check spectra of different resolution and instruments.

The spectral package can also read ASCII files allowing the user to import his/her own data to compare with the data accessed from remote archives.

In addition, the spectral application can start external specialized analysis packages such as XSPEC (Shafer *et al.* 1991).

3.4   THE TIMING APPLICATION

The ESIS Timing application allows the user to retrieve, display and manipulate time series data. Like all other data manipulation packages, the timing application has tools to browse, retrieve and manipulate data from the EXOSAT, *Einstein* and Rosat databases. The timing application is made up of a graphical interface that receives input from the user and a general purpose timing analysis package called XRONOS (Stella & Angelini 1992) that executes the commands. The result of this process is generally a plot that is displayed using the Athena Widget Plotter package. The most important capabilities supported are

▷ Retrieve and display time series data,

▷ General statistical analysis,

▷ Power spectrum, and

▷ Auto correlation.

More detailed analysis can be performed by starting XRONOS in native mode. The user own data can also be read as ASCII or FITS files.

3.5   THE BIBLIOGRAPHY SERVICE

The ESIS bibliographic service, also known as ESISBIB, contains an extract of the NASA Abstracts file and all the bibliographic references from the SIMBAD database. Abstracts date back to 1962, whereas SIMBAD references date back to around 1950. There are currently about 500,000 references altogether. Each NASA reference has been carefully cross-correlated with SIMBAD references, to avoid redundancy.

The ESISBIB Graphical User Interface is a query form which allows the user to query multiple fields. Results may be saved in files, including the original queries. A unique feature of the database is the ability to use SIMBAD to search on the bibliographic references of a given object and retrieve not only the references, but also the abstracts. In recent months the project has made ESISBIB available through the World Wide Web. Many of the original functions have been implemented and new search and help capabilities have been introduced. The most prominent of those is the index-browse feature, which is a facility to help the user preparing a query.

Since ESISBIB is now available on the World-Wide Web ([→2]), it is possible *for any* author of an article to actually hyperlink reference lists to their actual abstracts. All the author needs to do is to query the bibliographic database for that given reference and to hyperlink the search-URL to the reference.

18–4   The ESIS Architecture

As we have discussed up to now, ESIS may be accessed in a variety of different ways. The main components of the actual ESIS system are:

CATALOGUES AND LOGS. A set of ORACLE tables, the ESIS catalogues and mission logs were chosen by the astronomical community, not only to be as general as possible (*e.g.*, photometric, astrometric measurements), but also to guarantee wide coverage of the electromagnetic spectrum (*e.g.*, flux values, magnitudes, etc.) The mission logs are used as pointers to remotely archived data products of all the major astronomical space missions.

SIMBAD. The Set of Identifications, Measurements and Bibliography for Astronomical Data (SIMBAD), provided by the Centre de Données astronomiques de Strasbourg (CDS) plays an important rôle in ESIS to resolve names, provide coordinates and bibliographic references of astronomical objects.

REMOTE ARCHIVES. ESIS cannot function properly without the interface to the remote archives. The remote archives provide ESIS with crucial data products. Today, the major remote archive sites integrated in ESIS are: VILSPA – IUE, ST-ECF, ESTEC, HEASARC and CADC.

REFERENCE DIRECTORY. In order to overcome the various naming conventions of fields in catalogues and the heterogeneity in units of measure used by the astronomical community's sub-components, it is necessary to homogenize access (see Ansari *et al.*, 1994b). The rôle of the ESIS Reference Directory is to provide the ESIS system with all the information necessary on each individual catalogue in order to guarantee uniformity. Identifiers (e.g. HD, HR, etc.) used in the various catalogues are also declared in the Reference Directory.

QUERY ENGINE. The Query Engine provides a set of functions that allow the programmer to make searches on ORACLE catalogues using SQL, as well as using a remote SIMBAD to resolve queries. The remote access to data products is treated externally, making the Query Engine library static, rather than implementing each time a new function to access new remote archives. The only prerequisite is that a catalogue with data product pointers (names of datasets, or an access method procedure) be made available to the Query Engine (Ansari *et al.*, 1994c).

ESISBIB. : Based on the Ful/Text documentation management system, ES-ISBIB has its own Applications Program Interface (API). This library of high-level modules simultaneously accesses three bibliographic *collections* (i.e., the NASA, SIMBAD, and cross-correlated collections) and retrieves references and abstracts. The API also makes use of a specially-built client/server that queries the SIMBAD database on any object and retrieves the reference code, which is then used to extract the full entries from the ESISBIB collections.

18–5   Conclusions

We have described in some detail the concepts behind ESIS, its architecture and the services available. The current operational system is the combination of three components: a) all the initial ideas as layed out in the mid-80's, b) input and feed-back from many scientists, various committees and working groups, and c) the strong technological evolution of all its components. The result is a modern service that gives on-line access to a large amount of multi-frequency data and that allows users to handle and compare astronomical data in a simple and unique way.

The ESIS software was developed primarily during 1992-1993, at a time when hardware and software technologies, network infrastructure and data formats all underwent major changes. The present software had to adapt to all these rapid changes during the development phase. The standard development procedure (involving the generation of user requirements, negotiation with industrial developers, development phase, testing and formal acceptance) could not be applied because the life cycle of such an approach is much longer than the time scale of evolution of many of the fundamental elements of ESIS. A mixed prototyping/formal approach was followed instead. This approach required the construction of an initial prototype by a small scientific group inside ESIS. This prototype was then "engineered", i.e. upgraded to high quality standards, by a group of software engineers working along with the scientific group.

The ESIS client and WWW software is freely available and can be retrieved from anonymous ftp at the address [→3]. Detailed and updated documentation about the system, together with examples and a cookbook are available on-line and can be directly accessed from the ESIS home page at [→1].

References

[1] Ansari, S.G.; Giommi, P.; Micol, A (1994a), ESIS on the World Wide Web, *Proceedings of the First Topical Seminar on World Wide Web and beyond for Physics Research*, Centro Studi "I Cappuccini", March 1994, San Miniato, Italy (in press).

[2] Ansari, S.G.; Giommi P.; Micol A.; Natile, P. (1994b), Homogenous Access to Data Reference Directory, in *Astronomical Data Analysis Software and Systems III*, D.R. Crabtree, R.J. Hanish, and Barnes, ed., ASP Conference Series, **61**, 139.

[3] Ansari, S.G., Giommi P., Stokke H., Preite-Martinez A. (1994c), Homogeneous Access Astronomical Data - The ESIS Experience, Proceedings to *Handling and Archiving Data from Ground-based Telescopes* M. Albrecht, F. Pasian ed., ESO Conference and Workshop Proceedings **50**, 143.

[4] Egret D., Wenger M., Dubois P. (1991), in "Databases & on-line Data in Astronomy", M. Albrecht and D. Egret ed., Astrophysics and Space Science Library **171**, 79.

[5] Giommi P., Angelini L., Jacobs P. and Tagliaferri G., (1992), in "Astronomical Data Analysis Software and Systems I", D.M.Worrall, C. Biemesderfer and J. Barnes ed., A.S.P. Conf. Ser. **25**, 100.

[6] Giommi P. and Ansari, S.G. (1994), The European Space Information System, in *Frontiers of Space and Ground-Based Astronomy*, W. Wamsteker, M.S. Longair, Y. Kondo ed.

[7] Giommi P., White N.E., and Angelini, (1995), in *Astronomical Data Analysis Software and Systems IV*, H. Payne, and J. Hayes, ed., ASP Conference Series, in press.

[8] Klingebiel, P., (1992), Using The AthenaTools Plotter Widget Set. Univ. of Paderborn, Germany.

[9] Kurucz, R.L, Peytremann, E. (1975), *SAO Special Report* **362**.

[10] Shafer R., et al. (1991), *XSPEC User's Guide*, ESA TM-9.

[11] Stella L. and Angelini L. (1992), in *Astronomical Data Analysis Software and Systems I*, D.M.Worrall, C. Biemesderfer and J. Barnes ed., A.S.P. Conf. Ser. **25**, 100.

[12] Wamsteker, W., Driessen, C., Munoz, J. R., Hassall, B. J. M., Pasian, F., Barylak, M., Russo, G., Egret, D., Murray, J., Talavera, A., Heck, A. (1989), *Astron. Astrophys. Suppl.* **79**, 1.

Access Pointers

[→1] ESIS Home Page: http://www.esrin.esa.it/htdocs/esis/esis.html
[→2] ESISBIB: http://www.esrin.esa.it/htdocs/esis/esisbib.html
[→3] ESIS client and WWW software:
 ftp://mesis.esrin.esa.it/pub/esis/software/

19

The Star*s Family: An example of comprehensive yellow-page services

André Heck

19–1   Introduction

In the book *Databases & On-line Data in Astronomy* (Albrecht & Egret 1991), the chapter entitled *Astronomical Directories* (Heck, 1991) was essentially dealing with classical publications on paper: directories of organizations, lists of electronic addresses, and so on. Some files started to be retrievable electronically. In Sect. 8 (*Future Trends*) of that chapter, we were looking forward to using on-line and networking facilities that have become a daily reality since.

Meanwhile indeed, what is called the information technology (IT) revolution or, as we prefer to label it, the IT *evolution* (cf. Heck, 1995, in this book, chapter 1) has brought major modifications in the way information is handled with new techniques (hardware, connectivity, *etc.*) and new tools (client/server facilities, resource discovery packages, hypertext/hypermedia concepts, *etc.*). The existing reference products tuned themselves to the new capabilities and new ones naturally derived from the new media available (this is typically the case for *AstroWeb* detailed below).

Compared with Albrecht & Egret 1991, the present book is a striking example of this evolution. Refer also to Heck & Murtagh (1993) for a review on advanced information retrieval tools with special emphasis on astronomy and related space sciences.

The concept of *yellow-page services* was probably publicly introduced for the first time in our community at the *ALD-II Conference* in Haguenau (Heck & Murtagh, 1992). The following lines will attempt to give a flavor of what can be achieved currently and will exemplify what a comprehensive set of products, the *Star*s Family*, can offer nowadays. Please refer to Heck (1991) for a historical review of the earlier directories and of our own compilatory activities that started some twenty years ago.

D. Egret and M. A. Albrecht (eds.), Information & On-Line Data in Astronomy, 195–205.

19–2   An overview of the Star*s Family

The *Star*s Family* is a growing collection of directories, dictionaries, databases and more generally on-line yellow-page services. The products are organized essentially around three sets of master files:

  ▷ the directory *StarGuides* (Heck, 1993a and 1994b) of astronomy, space sciences, and related organizations of the world, with its associated database *StarWorlds* (Heck *et al.*, 1994);

  ▷ the dictionary *StarBriefs* (Heck, 1993b and1994a) of abbreviations, acronyms, contractions, and symbols in astronomy, space sciences, and related fields, with its associated database *StarBits* (Heck *et al.*, 1994);

  ▷ the database *StarHeads* (Heck, 1994c) of individual web pages essentially of astronomers and related space scientists.

The publications on paper (*StarGuides* and *StarBriefs*) are available at request as *CDS Special Publications* (refer to the bibliography) and are basically updated monthly. The databases have been made available at the Centre de Données astronomiques de Strasbourg (CDS) and are reachable through the CDS WorldWideWeb (White, 1993) server accessible *i.a.* via Mosaic (Hardin, 1994). The databases are updated 'permanently'. As we shall see in the following section, the compilation of all the master files is fortunately synergetic.

19–3   Scope of the master files

The directory *StarGuides* is the new name of the directory *Astronomy, Space Sciences and Related Organizations of the World (ASpScROW)* (Heck, 1990), itself resulting from the merging and scope broadening of the earlier directories IDPAI (*International Directory of Professional Astronomical Institutions* – see *e.g.* Heck, 1989b) and IDAAS (*International Directory of Astronomical Associations and Societies* – see *e.g.* Heck, 1989a), both subtitled 'together with related items of interest'.

StarGuides gathers together all practical data available on associations, societies, scientific committees, agencies, companies, institutions, universities, etc., more generally organizations, involved in astronomy and space sciences. Many other types of entries have also been included such as academies, bibliographical services, data centres, dealers, distributors, funding organizations, IAU-adhering organizations, journals, manufacturers, meteorological services, national norms and standards institutes, parent associations and societies, publishers, software producers and distributors, and so on.

Besides astronomy and related space sciences, other fields such as aeronautics, aeronomy, astronautics, atmospheric sciences, chemistry, communications, computer sciences, data processing, education, electronics, engineering, energetics, environment, geodesy, geophysics, information handling, management, mathematics, meteorology, optics, physics, remote sensing, and so on, are also covered when appropriate.

Currently more than 5000 entries from about 100 countries have been selected. The information is given in an uncoded way for easy and direct use. For each entry, all practical data available are listed: city, postal and electronic–mail addresses; telephone and telefax numbers; Uniform Resource Locators (URL) for on–line services; foundation years; numbers of members and/or staff; main activities; titles, frequencies, ISS Numbers and circulations of periodicals produced; names and geographical coordinates of observing sites; names of planetariums; awards, prizes or distinctions granted; and so on.

The basic philosophy of these directories is to provide practical data which one seeks always to have at one's disposal. They have proved over the years to be not only valuable auxiliaries for improving national and international relationships, but also efficient tools for helping laypersons and public bodies to contact organizations easily.

As to the dictionary *StarBriefs*, it derives from a list of acronyms that was included in the 1990 editions of IDAAS and IDPAI, but it had become so voluminous when compiling ASpScROW 1991 that it had become more appropriate to provide it as a separate, nevertheless complementary, publication (Heck, 1990a). It gathers together presently more than 75,000 abbreviations, acronyms, contractions and symbols with sections devoted to Greek letters, mathematical symbols, special signs and characters, as well as to entries with a numerical beginning. The field coverage is similar to that of the directories.

Abbreviations, acronyms, contractions and symbols in common use and/or of general interest have also been included when appropriate. The travelling scientist has not been forgotten (codes of airlines, locations, currencies, and so on) and humour is not quite absent from this publication either.

The underlying idea was to offer to astronomers and space scientists a practical assistant in decoding the numerous abbreviations, acronyms, contractions and symbols that they might encounter in their range of activities, including travelling. Maybe a bit paradoxically, if scientists quickly grasp the meaning of an acronym purely in their specific field, they will probably have more difficulties with adjacent fields. It is actually for this purpose that this dictionary might be more often used. Scientists might also use this compilation to avoid assigning an acronym that has already too many or confusing meanings.

This compilation is essentially carried out in parallel with the permanent updating of the directory *StarGuides*. In practice, all major abbreviations and acronyms encountered when scanning the general literature and the documentation received in relation with the directory are gathered, the underlying principle being that they might also appear one day under the eyes of astronomers and space scientists.

StarWorlds is the database version of the directory *StarGuides* and *StarBits*, the database version of the dictionary *StarBriefs*. The URLs provided in the database of organizations *StarWorlds* are actually active items that can be used through tools such as Mosaic to move onto (or 'navigate' towards) the corresponding documents. These WWW 'pages' are themselves often linked to web 'pages' of individuals, themselves in turn pointing to other resources. The database *StarHeads* gathers together the URLs of such WWW 'pages' of individuals. At the time of writing, only a few months after starting the compilation, more than 500 items are accessible. But this figure is increasing weekly and it would be interesting to see how big will be the database by the time this chapter will be read.

19–4   Accessing the information

The databases are directly accessible via Internet and tools such as Mosaic through the following Uniform Resource Locators (URLs):

▷ for StarWorlds: `http://cdsweb.u-strasbg.fr/~heck/sfworlds.htm`

▷ for StarBits: `http://cdsweb.u-strasbg.fr/~heck/sfbits.htm`

▷ for StarHeads: `http://cdsweb.u-strasbg.fr/~heck/sfheads.htm`

All together, they can be reached via the CDS homepage [→1] giving access to the various CDS services, as well as to external astronomy resources such as *AstroWeb* (see below).

The hypertextual structure of the databases on the CDS Mosaic server includes, beyond the search mechanisms, general introductory documents, access to forms, tips for usage, hot news, e-mailing facilities, lists of national telephone, telefax and telex codes, and so on. All the facilities cross-point to each other, with, as already indicated earlier, possible active navigation via retrieved links. At the time of writing, upgrading plans include some logical syntax capabilities, underlying thesaurus structure and, last but not least, retrieval of observing facilities on the basis of their location on our planet (especially useful for observing campaigns).

For users or libraries who would prefer the three-dimensional and stereoscopic browsing through a publication on paper, the directories are provided with an exhaustive index giving a breakdown not only by different designations and acronyms, but also by location and major terms in names. Quite a few thematic subindices are also provided as well as a list of telephone and telefax national codes.

19–5   Other Star*s Family products

StarWorlds and *StarBits* have not been the only databases set up from the *Star*s Family* master files. The very first database was called *StarWays* (Heck et al., 1992) and had been implemented by the ESIS group at ESRIN, an establishment of the European Space Agency (ESA) located at Frascati, Italy. There are plans to reactivate it, depending of course of ESIS' future.

The European Southern Observatory (ESO) located in Garching, Germany, has set up the databases *StarGates* and *StarWords* (Albrecht & Heck, 1993, 1994a&b) respectively associated to the directory *StarGuides* and to the dictionary *StarBriefs*. They are accessible through the standard Starcat account. Although these databases are used internally, the ESO Mosaic pages are now pointing towards the CDS databases *StarWorlds* and *StarBits* described earlier.

The *Star*s ·Family* line encompasses also other types of products such as sets of mailing stickers (*StarLabels*) made available at production cost to meeting organizers, as well as to publishers, manufacturers, and so on under some conditions of requirement and usage. The same applies to *StarSets*, tailored subsets of the master files on various media.

19–6   Maintenance and quality

The *Star*s Family* compilations · have taken advantage of the experience gained with each successive release, especially in the development of techniques for collecting, verifying and treating the data. To compile a directory or a database of real value is indeed quite a different venture to just reproducing and distributing, with comments of greater or lesser interest, data collected indiscriminately from all available sources. If professional file construction techniques are necessary, they cannot save the extensive background, unrewarding and very careful work which is indispensable for the compilation of a valuable resource.

The definition of a very well profiled and adapted questionnaire, the homogenization of the data collected and the maximum reduction of the respondents' biases are all points that must be satisfied, often with the help of the most modern communication means. The continuous political evolution of the world has also to be taken into account. If the information is provided in the *Star*s Family* files 'bone fide', the best is however done to keep track of the modifications happening and to implement them as soon as they are confirmed or recognized by the international community.

One could never stress enough the importance of this obscure daily work consisting of patiently collecting data, checking information and updating the master files. If scientists have a natural tendency to design projects and software packages involving the most advanced techniques and tools, there is in general less enthusiasm for the painstaking and meticulous long-term maintenance which builds up however the real substance of the databases. This has also to be carried out by knowledgeable scientists or documentalists and cannot be delegated to unexperienced clerks.

The fashion is now shifting towards designing and testing quality control processes, but we believe that the best quality assurance (accuracy, homogeneity, exhaustivity, *etc.*) has to be achieved when collecting and entering the data themselves. None of the algorithms currently available has really convinced us of its absolute necessity and satisfactory utility. Again here, developing such processes is an appealing challenge for scientists, but most of the algorithms designed work statistically. For a database user, it does not matter much whether it is accurate up to 90% or 95%. The user wants to find the piece of information he/she is looking for, and, if found, this has to be accurate. All these considerations are obvious if a phone book is taken as a model for yellow-page services.

19–7   Speaking of evolutions

The profile of the directories and the databases, as well as the questionnaires sent to the various organizations listed have been improved and adapted along the years. The information gathered has been more and more comprehensive. The categories of the entries listed have also been gradually broadened to better serve the needs of astronomers and space scientists. When compiling files such as the *Star*s Family* master ones, one cannot but be impressed by the very broad spectrum of disciplines to which astronomy and related space sciences are linked, and by the very large variety of techniques applied in these fields.

The successive releases of the directories and databases give fairly accurate global pictures of the active organizations in the fields covered. Their sequence testifies the sometimes rapid evolution of scientific interests, of data collecting and handling techniques, as well as of communications in the broad sense. A few countries have also rearranged the structure of their national facilities in the course of the past years.

Among the most recent striking changes, electronic mail has dramatically modified the way scientists communicate and exchange files of all kinds: a revolution in this era of communication similar to that of the computer earlier on. The way e-mail has appeared has been first as many addresses on different networks for only a few institutions. Then along the years, a few networks have emerged (as several of them also merged) and e-mail became widespread. Typically now, practically all organizations and many individuals are connected, but to a maximum of two networks only.

Telefax facilities have also become widespread. Telex and teletext are consequently becoming less fashionable and disappear gradually.

Of course, the political evolution in the world is directly reflected in the information provided. The USSR, the German Democratic Republic and Czechoslovakia have disappeared. New countries lengthened *StarGuides'* table of contents. The liberalization of political regimes, especially in Eastern Europe, has resulted in a dramatic increase of the questionnaires returned from the regions concerned. Most of the African continent remains however a dramatic gap (refer to the maps published in Heck, 1991, that remain largely valid).

Technical evolutions have also been playing a significant rôle in the recent years and we went through them while producing the successive versions of the directories and setting up the databases. Desktop and electronic publishing, with all the indexing facilities, have become instrumental for providing monthly updated releases of the *Star*s Family* products on paper with an outstanding quality essentially due to the TEX typesetting system.

As to the databases, not only flexible management systems allow efficient information retrieval, but the last couple of years have seen the explosion of WWW servers and of navigation tools such as Mosaic (refer to Heck, 1995 and to the references quoted therein). An on-line resource is nowadays identified by its URL and this led quite naturally to the compilation of databases of URLs ...

19–8   AstroWeb

AstroWeb (Jackson *et al.* , 1994) is a new WWW resource providing links to astronomy resources available on Internet and using *i.a.* the Mosaic client tool. Links are anchors or hypertext references to the listed resources through implicit URLs. These links allow transparent access to information provided by WWW, gopher (Anklesaria, 1993), anonymous ftp, WAIS (Fullton, 1993), Usenet News, and Telnet services. Thus *AstroWeb* may appear as the organizational counterpart of *StarHeads* devoted to individuals, but most of *AstroWeb*'s URLs are included in *StarWorlds* too, together with additional information.

The *AstroWeb* database is maintained by the members of a consortium located at five institutions: CDS, MSSSO, NRAO, ST-ECF and STScI. Each of these supports a different version of *AstroWeb* that have distinctive styles and contents, but all are computed from the same master database, which is coded in an agreed interchange format.

Each resource record is categorized and many resources have a paragraph describing the resource and containing links to other records. The database is queried on a regular basis to identify which URLs no longer work and the merged listing is edited to reflect these changes.

Astroweb is reachable through the following URLs:

▷ CDS site: `http://cdsweb.u-strasbg.fr/astroweb.html`

▷ MSSSO site: `http://meteor.anu.edu.au/anton/astronomy.html`

▷ NRAO site: `http://fits.cv.nrao.edu/www/astronomy.html`

▷ ST-ECF site: `http://ecf.hq.eso.org/astro-resources.html`

▷ STScI site: `http://stsci.edu/net-resources.html`

The *AstroWeb* database has been designed to facilitate distributed maintenance. Specific on-line facilities have been made available for this purpose. There are also on-line forms by which new resources can be added and existing resources can be edited.

19–9   A few last comments

The reader is invited to refer to the other chapter by the same author in the present book (Heck, 1995, page 1) for more general issues related to IT, database maintenance, quality control, and so on.

The development of navigation tools made it much easier to set up effective distributed information networks. As to yellow-page services, we can only encourage organizations (institutions, associations, funding agencies, and so on), as well as individuals, to set their own resources reachable on WWW. Comprehensive files of e-mail addresses such as Benn & Martin's (1990) ones should also be directly integrated, once a procedure will be designed to keep them efficiently up to date.

Here is a request and/or an advice to all producers of documents: the meaning of abbreviations, acronyms, contractions and symbols should always be explained at their first occurrence in any document. It would be even more practical and helpful if authors and editors systematically tabulated them with their explanations at the end of the papers, chapters, reports, proceedings, books, etc., they produce. Such a policy would at least remove any ambiguity, since, as shown many times in *StarBriefs* and *StarBits*, a given string of characters may have numerous meanings.

A non-negligible amount of time has been devoted to shape this paper (which is a quasi-linear document since the only breaks here are the bibliographical references) from fragmented information with physical or virtual hypertextual links, closer to the author's brain structure (the same is actually true for many persons). May the publisher produce the third edition of this book through hypertext/hypermedia means and thus save precious time!

Acknowledgements

It is a pleasure to acknowledge here the impact of numerous readings and conversations we do not dare detailing in order to avoid forgetting a couple of them whose contribution, if sometimes not always important in substance, has been nevertheless enlightening.

Special thanks are directed to all persons who assisted us over the years in the materialization of the *Star*s Family* products at all levels. Most of their names appear in the bibliography and in the quoted papers. A special mention is due here to Daniel Egret and François Ochsenbein for their collaboration in making the databases accessible on the CDS WWW server through efficient search mechanisms.

Last, but not least, we are very grateful to all persons and organizations who contributed to the very substance of the *Star*s Family* products by returning the questionnaires, by providing spontaneously the relevant documentation on-line, or otherwise. The *Star*s Family* products have been conceived for them and for the vast community of users. We are looking forward to satisfying their needs even better in the future.

References

[1] Albrecht, M.A. & Heck, A. 1993, StarGates and StarWords – An on-line yellow-page directory for astronomy, *ESO Messenger* **73**, 39-40

[2] Albrecht, M.A. & Heck, A. 1994a, StarWords – A database of abbreviations, acronyms and symbols in astronomy, space sciences and related fields (announcement of a database), *Astron. Astrophys. Suppl.* **103**, 471

[3] Albrecht, M.A. & Heck, A. 1994b, StarGates – A database of astronomy, space sciences and related organizations of the world (announcement of a database), *Astron. Astrophys. Suppl.* **103**, 473-474

[4] Anklesaria, F. & McCahill, M. 1993, *The Internet gopher*, in *Intelligent Information Retrieval: The Case of Astronomy and Related Space Sciences*, eds. A. Heck & F. Murtagh, Kluwer Acad. Publ., Dordrecht, pp. 119-125.

[5] Benn, Ch. & Martin, R. 1990, Electronic Mail Guide and Directory, Roy. Greenwich Obs., Herstmonceux (electronic files)

[6] Fullton, J. 1993, *WAIS*, in *Intelligent Information Retrieval: The Case of Astronomy and Related Space Sciences*, eds. A. Heck & F. Murtagh, Kluwer Acad. Publ., Dordrecht, pp. 113-117.

[7] Hardin, J. 1993, Human collaborations technologies for the Internet - NCSA Mosaic and NCSA Collage, communication to *Astronomical Data Analysis Software and Systems III*, Victoria, BC, 13-15 Oct. 1993

[8] Heck, A. 1989a, International Directory of Astronomical Associations and Societies together with related items of interest – IDAAS 1990, *Publ. Spéc. Centre Données Strasbourg* **13**, vi + 716 p. (ISSN 0764-9614 – ISBN 2-908064-11-1)

[9] Heck, A. 1989b, International Directory of Professional Astronomical Institutions together with related items of interest – IDPAI 1990, *Publ. Spéc. Centre Données Strasbourg* **14**, vi + 658 p. (ISSN 0764-9614 – ISBN 2-908064-12-X)

[10] Heck, A. 1990a, Acronyms and Abbreviations in Astronomy and Space Sciences, *Publ. Spéc. Centre Données Strasbourg* **15**, ii + 150 p. (ISSN 0764-9614 – ISBN 2-908064-13-8)

[11] Heck, A. 1990b, Astronomy, Space Sciences and Related Organizations of the World – ASpScROW 1991, *Publ. Spéc. Centre Données Strasbourg* **16**, x + 1182 pp. (ISSN 0764-9614 – ISBN 2-908064-14-6) (two volumes)

[12] Heck, A. 1991, Astronomical directories, in *Databases & On-Line Data in astronomy*, eds. M.A. Albrecht & D. Egret, Kluwer Acad. Publ., Dordrecht, pp. 211-224

[13] Heck, A. 1993a, StarGuides – A directory of astronomy, space sciences and related organizations of the world (announcement of a catalogue), *Astron. Astrophys. Suppl.* **102**, 85-86

[14] Heck, A. 1993b, StarBriefs – A dictionary of abbreviations, acronyms, and symbols in astronomy, space sciences, and related fields (announcemnt of a catalogue), *Astron. Astrophys. Suppl.* **102**, 87

[15] Heck, A. 1994a, StarBriefs 1994 – A dictionary of abbreviations, acronyms, and symbols in astronomy, space sciences and related fields *Publ. Spéc. CDS* **22** vi + 818 pp. (ISSN 0764-9614 – ISBN 2-908064-20-0)

[16] Heck, A. 1994b, StarGuides 1994 – A directory of astronomy, space sciences and related organizations of the world, *Publ. Spéc. CDS* **23** viii + 880 pp. (ISSN 0764-9614 – ISBN 2-908064-21-9)

[17] Heck, A. 1994c, StarHeads (announcement of a database), *Astron. Astrophys. Suppl.*, **109**, 265.

[18] Heck, A. 1995, Facets and challenges of the information technology evolution, this book, page 1.

[19] Heck, A., Ciarlo, A. & Stokke, H. 1992, StarWays – A database of astronomy, space sciences and related organizations of the world (announcement of a database), *Astron. Astrophys. Suppl.* **96**, 565-566

[20] Heck, A., Egret, D. & Ochsenbein, F. 1994, StarWorlds – StarBits (announcement of two databases), *Astron. Astrophys. Suppl.*, **108**, 447.

[21] Heck, A. & Murtagh, F. 1992, Astronomy from large databases II. Haguenau, 14-16 September 1992, *ESO Conf. & Worshop Proc.* **43**, x + 534 pp. (ISBN 3-923524-47-1)

[22] Heck, A. & Murtagh, F. 1993, Intelligent information retrieval: the case of astronomy and related space sciences, Kluwer Acad. Publ., Dordrecht, iv + 214 pp. (ISBN 0-7923-2295-9)

[23] Jackson, R., Wells, D., Adorf, H.M., Egret, D., Heck, A., Koekemoer, A. & Murtagh, F. 1994, AstroWeb – A database of links to astronomy resources (announcement of a database), *Astron. Astrophys. Suppl.*, **108**, 235.

[24] White, B. 1993, WorldWideWeb (WWW), in *Intelligent Information Retrieval: The Case of Astronomy and Related Space Sciences*, eds. A. Heck & F. Murtagh, Kluwer Acad. Publ., Dordrecht, pp. 127-133

Access Pointers

[→1] CDS Homepage: http://cdsweb.u-strasbg.fr/

[→2] StarWorlds: http://cdsweb.u-strasbg.fr/~heck/sfworlds.htm

[→3] StarBits: http://cdsweb.u-strasbg.fr/~heck/sfbits.htm

[→4] StarHeads: http://cdsweb.u-strasbg.fr/~heck/sfheads.htm

[→5] CDS site: http://cdsweb.u-strasbg.fr/astroweb.html

[→6] MSSSO site: http://meteor.anu.edu.au/anton/astronomy.html

[→7] NRAO site: http://fits.cv.nrao.edu/www/astronomy.html

[→8] ST-ECF site: http://ecf.hq.eso.org/astro-resources.html

[→9] STScI site: http://stsci.edu/net-resources.html

20

Library Information Services

U. Michold, M. Cummins, J. M. Watson, J. Holmquist &
R. Shobbrook

20–1   Introduction

Although more and more information is being made available over the Internet, for many subject areas it is still true that reliable bibliographic information can only be found in the various commercial databases. Fortunately, astronomy is an exception in this regard. Thanks to many national and international projects as well as to the personal initiative of astronomers, computer scientists, librarians and publishers, members of the astronomical community can make use of a variety of non-commercial bibliographic resources on the Internet.

As commercial databases have been described in detail before (see *e.g.* Watson 1991), this chapter will concentrate on Library information services that require neither an account nor a password and do not charge fees according to the connecting time or the number of data sets that are displayed or downloaded.

Bibliographical information is manyfold, and the categories mentioned below are overlapping. As with any other enquiry, when searching for bibliographic information it is useful to have an idea of what can be found where before one starts a search. The World-Wide Web (WWW) is an excellent tool, making access to bibliographical information as easy as currently possible. The AstroWeb, a list of over 1000 astronomical Web links, is the recommended entry point to find bibliographical resources on the Internet.

The idea of using only one single database to answer all kinds of information needs for bibliographical purposes will remain a dream for a while. However, it is hoped that access to the various databases will become even smoother and more unified.

D. Egret and M. A. Albrecht (eds.), Information & On-Line Data in Astronomy, 207–228.
© 1995 *Kluwer Academic Publishers.*

20-2   Library Catalogs and Holdings

No matter how far technology has brought us, we should bear in mind the original roots and principles of information storage and retrieval. Without the meticulous work of generations of librarians and their assistants, there would have been no catalog records to automate. During recent years, many astronomy libraries have converted their card catalogs into machine-readable form and made them accessible via the Internet. Their number is constantly growing. The list includes among others (in alphabetical order) the library of the Astronomical Observatory of Trieste (OAT) [→4], the European Southern Observatory (ESO) library [→15], the Fermilab library [→18], the Goddard library [→19], the library of the Harvard-Smithsonian Center for Astrophysics (CfA) (through the Harvard Online Information Server (HOLLIS) [→10], the Princeton Astrophysics library [→28], the library of the Royal Greenwich Observatory [→30], the library of the Space Telescope Science Institute (STScI) [→33], and, of special interest to the high-energy physics community, the Stanford Linear Accelerator Center (SLAC) library [→37]. At present, most libraries can be reached via a telnet link, although some of them make their holdings available directly on the Web through a searchable index.

Typically, these catalogs contain the title entries of the print and non-print media owned by a particular library. Additional information may include the availability of an item (on-loan/on-order *etc.*), contents tables *e.g.* of conference proceedings or new acquisitions lists. Depending on the library's policy, users can search for authors, titles, keywords, publisher and publication years and/or a combination of these. Therefore, library card catalogs on the Internet will be used to check the holdings of a particular library, to verify bibliographical references, or as suggestions for acquisitions.

In addition to libraries of particular institutes, the holdings of university, state and public libraries can be found on the Internet. Users should be aware of the fact that large libraries usually build their on-line catalogs by adding recently bought titles, whereas dated material might not yet be included. Probably the most frequently used general on-line library catalog is the Library of Congress Information System (LOCIS). It provides access to English language books from 1968 onward (foreign language books from approx. 1973), serials cataloged since 1973 as well as selected earlier and other material like maps, audiovisual items *etc.*No password is required, but the system can be unavailable depending on the opening hours of the Library of Congress [→22].

While library (book) holdings still have to be checked one by one, librarians have gone beyond this limitation with regard to journals —they created the Union List of Astronomy Serials (ULAS). The list, one of the ongoing projects of the Physics-Astronomy-Mathematics Division of the Special Libraries Association, was compiled by Judy Bausch of the Yerkes Observatory. It indicates which astronomy libraries own which volumes of journals and observatory publications and thus facilitates Inter-Library Loan and helps to locate difficult-to-find items. ULAS can be accessed on-line using a searchable index [→40].

Since it is difficult to monitor the rapidly changing situation of library catalogs on the Internet, colleagues volunteered to set up WWW pages, providing links to multi-disciplinary library catalogs around the world which can be accessed remotely [→44], [→45]. However, for astronomy the As-troWeb [→1], listing many publication-related resources, remains the first choice (Jackson et al. 1994).

Appropriate sources for the verification of bibliographic references or price information are also bookshops on the Internet [→72], [→73] as well as the various Web pages of publishers [→60]-[→71]. Most publishers have not yet made available their complete catalogs, but computer science, engineering, artificial intelligence, and chemistry are typically among the first areas to go on-line.

20–3   Indexing and Abstract Services

The topic of abstract services is the one where it becomes most obvious that astronomy is in a privileged situation compared to many other sciences: the astronomy community can obtain bibliographic information about individual published papers (as opposed to book titles) from various services and databases. These abstract services are not only quicker than their printed counterparts, many of them also make use of advanced (or "intelligent") information retrieval systems, including relevance feedback, hypermedia and direct-manipulation user interfaces (see *e.g.* Fullton 1993, Van Herwijnen 1993 and Adorf 1993).

Following are brief descriptions of some of the indexing and abstract services currently available on the Internet. For more details, please see the individual chapters in this book. Bibliographic information provided by commercial database hosts is dealt with below.

ADS: The Abstract Service of the NASA Astrophysics Data System (ADS) provides access to the NASA/STI abstracts in astronomy and related fields. Users can search for authors, SIMBAD object names, NASA/STI keywords, as well as any word in the title and in the abstract text of all major and many minor astronomical journals, NASA reports, and many Ph.D. theses. Based on already retrieved hits, an improved search can be formulated, and even the settings can be defined by assigning relative weights to particular characteristics [→2].

STELAR: is "a pilot project designed to study the technical and practical aspects of making the refereed scientific literature available on
line" (Warnock 1994). Making use of a WAIS (Wide Area Information
Servers) index, articles from all major journals can be found. The system also provides a searchable index to the abstracts of AAS meetings
since June 1992. It was planned to include scanned images of journal
pages, but because of financial problems, the future of STELAR is more
than uncertain. The journal issues already included in the STELAR
database can still be searched through [→39].

ESIS: The European Space Information System (ESIS), located at ESRIN in
Frascati, Italy, is a service that aims mainly at the space science community. The bibliographic service ESISBIB supports user queries by
authors, title words and words from the abstract. It is planned to
include SIMBAD object search in the next release (current release is
version 2.5) [→14].

SIMBAD: The SIMBAD (Set of Identifications, Measurements, and Bibliography for Astronomical Data) database allows user queries for basic
data, cross-identifications, observational measurements, and bibliography, for celestial objects outside the solar system. To open a SIMBAD
session, you have to register first in order to get a user identification
number. SIMBAD users will be charged for their searches [→31].

CDS: The Centre de Données astronomiques de Strasbourg (CDS) makes
available the abstracts from Astronomy & Astrophysics main journal
and Astronomy & Astrophysics Supplement Series as well as the Publications of the Astronomical Society of the Pacific (PASP) approximately
four weeks *before* their actual publication through the World Wide Web.
This service started with volume 284 (April 1994) of the A & A main
journal and volume 103 (January 1994) of the Supplement Series [→8].
For PASP, volume 106, number 695 (January 1994) is the first volume to
be included [→9]. The databases are searchable by keyword or authors'
names. Further, the contents tables of individual issues can be browsed
and individual abstracts viewed. CDS also provides access to the A &
A and A & A Supplement Series abstracts through an anonymous ftp
account [→8]. For PASP, this service is provided by the STScI [→33].

CfA: The Center for Astrophysics (CfA) offers a program that displays the
titles, authors and bibliographic references of papers published in the
Astrophysical Journal (Parts 1 and 2 and Supplements), in the Astronomical Journal, and in the Publications of the Astronomical Society
of the Pacific from January 1988 onwards [→11]. Users can specify a
publication year and browse the author or subject index. As a further
service, CfA provides access to the abstracts of papers accepted by but
not yet published in the ApJ Letters [→12]. The list is in reverse order

of acceptance so that the most recent papers appear first. These are open accounts, which require no password. They can be accessed by typing "username: APJAJ or APJLETT". Copyright to these abstracts remains with the American Astronomical Society (AAS), and permission to publish or distribute the material must be obtained from them.

NED: Bibliographic information regarding the subject field of extragalactic objects can be obtained from the NASA/IPAC Extragalactic Database (NED), where authors, titles, and abstracts of published papers from many astronomical journals as well as Ph.D. theses are included [→24].

UNCOVER: In addition to serving as a document delivery facility, the UnCover database also may be a useful resource for obtaining bibliographic information about published articles (although without abstracts). The database includes all major astronomical journals. Users need not set up an account, but may search the entire database for authors' names, words, or journal titles, as an Open Access user. Due to the concept of UnCover, the time between the publication of an article and its inclusion into the database is extremely short, since the libraries pertaining to this service enter bibliographical information upon receipt of a new journal issue and thus make the contents tables accessible for everybody at a very early stage [→41].

20–4 Preprint Services

Between the writing of a scientific paper and its actual publication a considerable delay occurs. In order to make the investigations of their scientists known to the community as early as possible, most astronomy institutes have been sending out paper preprints. All major astronomy libraries (plus many individual scientists) receive preprints from other astronomical institutes which they make available to their users, typically by displaying them for a certain time. According to a survey by Joyce Watson dating from March 1994 (referred to in Stevens-Rayburn 1994), many libraries enter the bibliographic data from the paper preprints they received into a database, allowing access for in-house scientists as well as sometimes for the public. Preprints are usually kept for between 3 and 24 months before being discarded.

This situation currently is undergoing major changes. In an effort to cut back their printing and mailing costs, institutes are looking for cheaper ways to distribute their preprints. Unfortunately, most institutes have taken individual decisions on how to do so and thus created a very inhomogeneous situation that makes finding preprints a tricky job. The large number of databases and servers with individual addresses and access procedures makes a search for preprints much more time-consuming and inefficient than it need to be.

Most institutes make their preprints available on a local server which can be accessed remotely by ftp, telnet, or through the World Wide Web. Depending on the institute's policy, a searchable index to their preprints is provided or they may simply be in chronological order. To alert the community of this service, some institutes send out lists of new preprints either in printed form or electronically. Printed copies often have to be requested directly from the author.

A few examples: Abstracts of recent STScI authored preprints can be found on a gopher [→36]. Preprints by staff and visitors of the National Radio Astronomy Observatory (NRAO) are available on the WWW as ASCII text abstracts plus (in most cases) a postscript version of the entire preprint [→27]. (NRAO and STSci databases of preprints from other institutes are mentioned below). The European Southern Observatory (ESO) provides access to the full text of many ESO preprints and some other publications [→17]. Abstracts of refereed staff publications from Canada-France-Hawaii Telescope (CFHT) can be viewed on the WWW [→6]. Four Italian institutes (Bologna Astronomical Observatory (OAB), Astronomy Department of Bologna University (DDA), Radioastronomy Institute of CNR (IRA), and TESRE Institute of CNR (ITE)) give access to a chronological listing of their preprints (plus bitmapped images of the abstracts) through the Bologna Astronomy Preprints service [→5]. The University of Massachusetts provides a list of preprints on star formation and from their Theory Group [→43]. Princeton Observatory preprints can be retrieved in electronic format (usually ASCII abstract and postscript full-text) [→29], and also the University of Amsterdam set up a server for electronic preprints and articles from their institute (currently scanned images) [→42]. For the subject area of high-energy physics, SLAC allows remote access to their HEP preprint database [→38], as does CERN through its Physics Information Service [→13]. For a more complete and up-to-date listing, please search the AstroWeb [→1].

In a paper presented at a workshop on electronic preprints, held in September 1994 in Baltimore, Sarah Stevens-Rayburn, librarian at STScI, pointed out that obviously "it is *really* important to institutions that their preprints be identified as *their* preprints. To have their preprints appear on some anonymous server is not good enough" (Stevens-Rayburn 1994).

An example of a different approach is the automated distribution service of the astro-ph archive at Los Alamos National Laboratory [→23] or its mirror site at SISSA, Trieste [→32], from which preprints from many participating institutes are forwarded to the community via e-mail. This e-print server also provides an archive from where preprints can be retrieved through a searchable index.

For librarians, these changes mean a considerable re-examination and modification of their rôle vis-a-vis preprints. Will they continue to request and keep at least one printed copy of each preprint for reference? Do they keep track of the addresses of the various servers where preprints can be obtained from and offer access to these servers as a standard library service? Do they just provide their users with instructions on how to ftp or telnet into a remote host and let them get the stuff they need themselves? Will they be the ones who finally try to facilitate access to preprints by setting up searchable databases containing the full-text of preprints from more than only one institution?

In addition, two important further questions come to mind: For how long should libraries keep electronic preprints in their databases, and should librarians add the full bibliographic reference to the preprint entry once the paper has been published?

Years ago, the latter question was "answered" by an initiative of Ellen Bouton and Sarah Stevens-Rayburn, librarians at NRAO, Charlottesville, and STScI, Baltimore, respectively. They provide the astronomical community with preprint databases to which full references are added upon publication of the paper- RAP Sheets and STEPsheets. The NRAO RAP Sheets [→26] are a bibliographic listing of all preprints received in the Charlottesville library of the National Radio Astronomy Observatory from 1986 forward. The database is indexed for searching via WAIS. The preprints include papers submitted to scientific and technical journals as well as those to appear in meeting proceedings. Correspondingly, the STScI STEPsheets [→34] are a list of preprints received at the Space Telescope Science Institute library during the last two years, including papers submitted to scientific and technical journals and to appear in meeting proceeding as well as all HST papers in the refereed literature. For preprints received at the STScI prior to the last two years (from 1982 onward), the STScI-OLDSTEP database can be searched [→35]. These databases are an invaluable source of information since they bridge the gap between when a paper is published and when it is included in any of the abstract services.

Institutes seem to continue creating individual databases instead of making a common effort to facilitate access to preprints. Librarians, on the other hand, already have developed ideas on what the ideal preprint control might be. Sarah Stevens-Rayburn characterized a desirable system as follows (Stevens-Rayburn 1994):

> "To make a preprint control system that is too complex "..." is to exacerbate the chaos and not help to control it. The ideal therefore includes the following:
>
> ▷ Ease of access: this means it should have an almost transpar-

ent user interface; no arcane command structures or multiple screens needed to get to the information.

▷ Support for the computationally disadvantaged. By this I *don't* mean no GUI interfaces or sophisticated forms. I *do* mean that there should also be a way for those without the equipment to deal with such to be able, for example, to send an electronic mail query to an information provider who will quickly search for the needed information and return it via electronic mail to the requestor. This service would, I hope, be totally automated, somewhat akin to `listserv` queries, and should cost the user no more per access than it costs the person with sophisticated equipment to make interactive queries.

▷ Inclusion of all authors, titles, subtitles, and, when it becomes available, publication information. If we are providing full text, this is obvious, but if we provide only a subset of full text, *i.e.* an index, complete information is important.

▷ Centrality of the indexing. What I mean by this is that it's perfectly fine for each institution to mount its own preprints, but the access to those preprints must be in one searchable database. There must also be a way for those people at institutions that do not have a central system to have their papers included."

Of course, people and organizations other than librarians have a vested interest in these questions. Publishers of the major astronomy journals have been remarkably lenient in turning a blind eye to the practise of distributing preprints. However, with the advent of wide distribution of preprints, by electronic means, and in such an organized, collective way, publishers are beginning to ask themselves about copyrights. Where does informal preprint distribution end and electronic "publication" begin? One way of solving this problem is to have the preprint text deleted upon publication. A reference to the published paper could be substituted.

In astronomy, the acceptance rate in the main journals is very high and the refereeing can often (some say "usually") result in major changes in the content of the article. It is important, therefore, to distinguish between the preprint and the final, published article. This is not so much the case in physics where the acceptance rate is lower.

Organizing databases to link to each other, maintaining references to published papers, indexing preprints, providing various front ends for various platforms, are all activities that require enormous financial and human resources. Whether these resources will be applied to ephemeral materials that consist of unrefereed science is a question that remains to be answered. Part of the answer lies in how quickly and effectively electronic journals are developed.

20–5   Electronic Journals and Newsletters

There are two distinct sides to electronic publishing of traditional journals. On the one hand there is the production process including submission, refereeing, editing, typesetting, printing. The other side is distribution. The enhancements to this latter, such as elaborate searching, access and display capabilities, are important benefits. Many subscribers, especially librarians, forget about the former aspect when they think of a journal as "going electronic". Yet it is here that the most progress has been made recently. Furthermore, the digitization of these processes is an essential step in the development of a completely electronic journal.

Many, if not all journals, are now accepting submissions in machine-readable form-compuscripts. Compuscripts are usually formatted with a LaTeX template which is distributed by the journal. The nature of these templates and what the journals do with the compuscripts varies. Astronomy & Astrophysics (through Springer-Verlag) uses a template that virtually forces authors to typeset their submissions. AAS, on the other hand, uses a template that is then converted to SGML for typesetting by the publisher. These different approaches have different advantages and disadvantages. Furthermore, even though many journals use the electronic submission for production, it appears that they still require a paper version for refereeing and editing. The remaining problem relating to compuscripts is the artwork —the technical problems relating to equations and tables have been solved, but figures and photos must still be submitted in camera-ready form. It is important to realize that compuscripts must still be screened for errors and edited, making estimates of cost savings difficult at this transitional stage. What is clear is that electronic processing of journal articles will save time between acceptance (i.e. post-refereeing) and publication. Astronomy and Astrophysics, for example, which receives about 80% of submissions electronically, has reduced that time by about three months.

The STELAR project (STudy of Electronic Literature for Astronomical Research) [→39] was an experiment by NASA and AAS to scan into digital form, much of the current (five years) major astronomical literature in order to assess the feasibility of making this literature electronically accessible to the astronomical community. It appears that, at least in the foreseeable future, the full text of the recent literature will not be available electronically.

No peer reviewed astronomy journal is distributed electronically. Some of the benefits of electronic distribution are nevertheless available because of the fact that the production processes have been digitized. Abstracts of the AAS meetings [→3], published in the AAS Bulletin, are available before the paper edition comes out, for example. So are abstracts for the Astrophysical Journal Letters (for instructions on how to obtain them, please refer to the section on *Recent Letters Abstracts* in the back cover of the ApJ Letters) as well as Astronomy and Astrophysics and its supplement [→8].

Electronic distribution and archiving of large sets of data is being done —Astronomy & Astrophysics makes them available through Strasbourg's CDS, AAS distributes a series of CD-ROM's that holds large tables *etc.*from articles in several journals. There is a new journal about to come out (The Journal of Astronomical Data, Twin Press) which will be *only* on CD-ROM and will contain large data sets.

AAS is on the brink of distributing an electronic version of the Astrophysical Journal Letters. They hope for full-scale distribution in the fall of 1995. A new set of articles and information about their electronic publishing program is available through the WWW [→3].

There are many issues relating to electronic journals- copyright, integrity of compuscripts, cost recovery, archiving (digital archiving is expensive and volatile), network crowding, rôle of libraries *etc.*that have not been resolved by AAS in their experimenting, nor by anyone else in the astronomical community. Therefore, paper will be the medium of choice for libraries for the foreseeable future, though not forever. Electronic distibution will probably be viable only for individual subscribers.

All the above is applicable to the traditional journals. There is a whole other set of publications, however, for which the same things do not apply —newsletters. There are more than a dozen of these [→46]-[→59], all with somewhat different purposes and approaches. Some come only in electronic form and started up that way, others started with print and now come in either or both formats. Many of them contain only abstracts collected on a certain subject specialty, others carry actual research news and observations. How they are handled by libraries varies also. Some distribute or archive them electronically, some print them out and archive them that way, some don't keep them at all. A problem for libraries is those newsletters which are not distributed but only made available on their local site. One editor of an association newsletter (Cassiopeia) is experimenting with sending out the table of contents electronically and providing access to the full text article at the local site.

20–6   Alerting Services

The most obvious alerting services that librarians provide are weekly lists of journal citations published on topics chosen by the researchers. An example of an automated service of this kind is "Current Contents with Abstracts" (electronic version on diskette described *e.g.* by Allen 1994 and Adorf 1993). Subscribers receive a weekly update with contents tables and abstracts from recently published journal issues. The program allows users to browse contents tables by issue, conduct searches based on various categories, and set up and save search profiles which can be run against new Current Contents issues.

But how current are these services? In our experience, the contents lag about three weeks behind receipt of the actual journal issues in the libraries. Nevertheless, researchers appreciate this service to assure themselves that they are not missing articles of interest.

More timely alternatives are services that routinely send tables-of-contents of previously selected journals as soon as new issues are published. One example is UnCover's Reveal service. As mentioned before, UnCover is an on-line document delivery service, *i.e.* users can search a database of the contents tables of nearly 17,000 periodicals including the major astronomical ones, select articles and order them from UnCover. Unlike other delivery services, UnCover also offers customers the possibility of subscribing to the alerting service UnCover Reveal. Users may choose journal titles of interest to them and will receive the full table-of-contents by e-mail whenever a new issue is entered into the database. It should be emphasized that there is no obligation for the user to actually order any article [→41].

A second example is the Springer-Verlag Journals Preview Service (SVJPS). For selected journals (mainly in the life sciences and radiology, but also including *e.g.* Communications in Mathematical Physics and Zeitschrift für Physik, A-D), tables-of-contents are sent to subscribers by e-mail approximately 10 days *before* shipping of a new issue. This service is free of charge [→69].

Nowadays, researchers in several areas of physics, including astrophysics, are alerted to new ideas in their fields by means of "preprint bulletin boards" like the e-print servers at SISSA [→32] and at Los Alamos [→23]. Preprints are not a new idea. Authors have distributed copies of their papers - pre-publication - since time immemorial. It is only with the advent of the electronic networks that such a wide and rapid distribution has become possible, forcing librarians and publishers alike to rethink their rôles and procedures (Stix 1994). It is important to note that all these preprints are unrefereed and unedited articles.

Librarians do not limit their alerting services to bibliographic information.

Keeping abreast of what's happening is essential to us, because we see our rôle as not only *retrieving* astronomical information, but also facilitating its *dissemination*. Various organizations and publishers have begun to include librarians in their discussions, *e.g.* the AAS Working Group on Astronomical Software, the AAS Publications Committee, and Commission 5 of the IAU. Information gained from this involvement is distributed *e.g.* on Astrolib and PAMnet (see "Informal Networking" section below), through which we alert each other (and the library users) to new developments.

A different kind of alerting service has been initiated by Liz Bryson, librarian at the Canada France Hawai Telescope (CFHT). She maintains a list of international astronomy meetings worldwide, a service which is heavily used by astronomers and librarians. Subscribers receive updates automatically through e-mail. The entire list can also be found on the WWW [→21].

20–7   Commercial Databases

Commercial databases cover every conceivable discipline, from basic science to medicine, business, banking, and any subject you care to name. We are fortunate in that astronomical literature has been covered for so many years by abstracting journals, going back to the Berliner Astronomisches Jahrbuch (varying titles, volume numbers, *etc.*) from 1776, and eventually evolving into the essential Astronomy and Astrophysics Abstracts that we rely on today from 1969.

The following is a short list of commercially available databases that may be accessed via computer, in most cases using an Internet connection. Please consult Watson 1991 for further information, telephone numbers, and contact names.

Vendors handle large numbers of databases, and individual contracts must be set up by libraries and organisations. Dialog, in Palo Alto, California, with international coverage, and STN International, in Karlsruhe, Germany, and gatewayed to the Chemical Abstracts Service in Columbus, Ohio, U.S.A., with offices in Japan and Australia are the largest Vendors. They take the individual databases from the publishers, and make them available to the users. In many of the relevant databases the cost per connect hour is high, but the coverage is good and timely, searching is straightforward and standardised, and training courses are available throughout the world.

> ▷ The Aerospace Database (Dialog File 108)
> This is the database of the American Institute of Aeronautics and Astronautics. It covers the open literature on all aspects of aerospace, including good coverage of astronomy and astrophysics. Until early 1994, the

Aerospace Database included material from the National Aeronautics and Space Administration Scientific and Technical Aerospace Reports (NASA/STAR), but these are now available separately (see below). (Print equivalent: International Aerospace Abstracts, monthly)

▷ Astronomy and Astrophysics Monthly Index
This is a privately-produced index available either on tape or in a printed version. Its purpose is to provide monthly updates to bridge the delay in publication of the Astronomy and Astrophysics Abstracts. It can be obtained from the publisher, Olivetree Associates, Sierra Madre, CA, and is not available via commercial vendors. (Printed equivalent: same title.)

▷ CONFSCI (Conferences on Energy, Physics and Mathematics; STN)
Lists conferences in above subjects, giving all information from first announcement, contacts, *etc.*, often some years ahead. However, does not include individual papers presented at meetings. (No print equivalent.)

▷ Conference Papers Index (CPI; Dialog File 77)
CPI complements CONF in that it covers the actual papers given at conferences, approximately 100,000 papers annually from 1,000 scientific meetings. (Print equivalent: Conference Papers Index.)

▷ Current Contents Search: Physical, Chemical and Earth Sciences (Dialog 440)
This is the on-line version of the weekly publication of the Institute for Scientific Information (ISI), which, literally, reproduces tables of contents of approximately 6,500 journals in all disciplines. (There are many different editions covering different disciplines.) As with many other databases, the vendors provide the facility for "saved searches." This enables the user to set up a personal profile, and access it frequently to obtain the latest information. (Print equivalent: Current Contents: Physical, Chemical and Earth Sciences)

▷ INSPEC (Information Services in Physics, Electrotechnology and Control; STN and Dialog File 2, 3, 4)
INSPEC covers the whole field of physics, plus editions covering the topics mentioned above. This is an extremely valuable database, and will become more so during the coming year, 1995.

The PHYS database, based upon the Physikalische Berichte, published by the Fachinformationszentrum Energie-Physik-Mathematik GmbH (FIZ) of Karlsruhe, Germany, will be discontinued in December, 1994. At this point, the database will be merged with INSPEC. This will undoubtedly show an increase in the coverage of astronomy and astrophysics, as FIZ has, for several years, concentrated on building up the coverage of these subjects. We hope, in the future, that agreements will be reached with the Astronomisches Rechen-Institut of Heidelberg for the incorporation of material from Astronomy and Astrophysics Abstracts. (Printed equivalent: Science Abstracts, Section A: Physics Abstracts)

▷ NASA/RECON

Until 1994, this database included the input from the International Aerospace Abstracts (q.v.). It is compiled of NASA reports and miscellaneous literature dealing with all aspects of aerospace research, including good coverage of astronomy and astrophysics. Access is made directly to the NASA Scientific and Technical Information (NASA/STI) in Linthicum, Maryland, who issue passwords, instructions, *etc.*It is available via the Internet. Many people now access this resource via the Astrophysics Data System (ADS) [→2], (see section on Indexing and Abstract Services in this chapter and separate chapter on ADS in this volume). The astronomical sections are conveniently searchable, and more sections will be added in future. (Print equivalent: NASA Scientific and Technical Aerospace Reports (STAR))

▷ National Technical Information Service (NTIS; Dialog file 6)

This is a U.S. government database, property of the U.S. Department of Commerce. It is extremely large and comprehensive, and covers all manner of government research reports, published and unpublished. It is available through all vendors, and comparatively very inexpensive. Its greatest value lies in in the coverage of the report literature on astronomy and astrophysics, as not only NASA, but the U.S. Air Force and Navy, including the Department of Defense, are seriously engaged in astronomical research. It is an interesting catchall, and always worth searching. It also covers the timespan from 1964 to date. (Print equivalent: has gone through many titles, since 1967 has been Government Reports Announcements.)

▷ SCISEARCH (Dialog files 34, 434; STN)

SCISEARCH is the multidisciplinary database operated by the Institute for Scientific Information, of Philadelphia, Pennsylvania. It contains many million records, and its value to us is somewhat different from those listed above. It is an invaluable source for cited references, where it is possible to track research back through time by articles that have made reference to earlier work. SCISEARCH became available in 1974, but one can trace articles written, in some cases, before the turn of the century, and list the references to them. It is also a good source for author searches which, combined with the cited reference capability, make the tracing of a scientist's work an easy matter. It is particularly favoured by those familiar with the printed version. (Printed equivalent: Science Citation Index.)

20–8   Informal Networking

Although a lot of bibliographical information can be retrieved through the Internet today, personal communication between librarians sometimes is quicker, more detailed and even more up-to-date and never should be underestimated. Astronomy librarians (and everybody interested in this subject area) can make use of two mailing listservs that were set up explicitly to cover their professional interests: Astrolib and PAMnet. Both lists are a forum for astronomy librarians, created in order to share information (including bibliographical references, reports on particular publications and the status of ongoing projects) and to prevent unnecessary duplication of work. Contrary to some other services on the Internet, all one needs to participate in these informal networking groups is an e-mail account.

The Astrolib mailing list is a moderated list.   Messages are collected and distributed by Ellen Bouton, librarian of the National Radio Astronomy Observatory (NRAO) in Charlottesville, Virginia.   To join Astrolib, send a message to library@annie.cv.nrao.edu

The PAMnet forum was initiated by members of the Physics-Astronomy-Mathematics (PAM) Division of the Special Libraries Association.   Many participants of PAMnet are in fact members of the SLA, but the list is open to everybody and does not oblige subscribers to become members. The list is managed by Joanne Goode of Miami University (Oxford, Ohio). PAMnet's address is sla-pam@listserv.lib.muohio.edu, Joanne Goode can be reached at goodejm@muohio.edu.

20–9   Use of a Thesaurus for On-Line Searching

Many users who formerly satisfied their information needs from their libraries (and librarian) are now accessing on-line databases for themselves. Librarians will continue to access databases on behalf of their clientele, however they will find they have a new rôle and an extension of service evolving: training of end-users (Fisher and Bjorner 1994), provision of current awareness search results, or just pointing users in the right direction with a login address. They will be expected to be familiar with the wide range of options available for finding information and may have to act as the intermediary on many occasions.   A thesaurus can be of valuable assistance to the subject novice be they the librarian, the student or the scientist in unfamiliar territory.   Ideally the thesaurus should be available in hardcopy and as an interactive on-line service. Lancaster 1972 referred to in Borgman, Moghdam and Corbett 1984 says "the thesaurus exists to lighten the intellectual burden on the searcher, to reduce the possibilities of human error, and to improve both the effectiveness and the cost-effectiveness of the search operation."

With the rapid increase in the availability of wide-area networking services there will need to be an assessment of the effectiveness of those on-line services and of the procedures involved in documenting and accessing the information. The commands and protocols between each database service sometimes vary, resulting in frustration and confusion. Librarians have been involved with the access and searching of (mostly commercial) databases for more than two decades and are only too familiar with some of the frustrations involved. Whilst some of the on-line resources are now becoming more user-friendly (Mosaic, WWW, WAIS) by the use of hypertext (clicking with a mouse on appropriate terms that serve as links to other documents), in some situations there is frequently the need to plan the search strategy via the use of keyterm expressions. A great deal of time, effort and sometimes money will be saved if the searcher plans the on-line search strategy *before* logging onto the database. This will apply to either numeric or bibliographical databases. Here reference will only be made to the on-line bibliographic search process and the method of access via *subject terminology*.

There are two approaches to subject searching, by natural language and by a controlled vocabulary *i.e.* a thesaurus. Boolean operators are normally used in conjunction with either approach. There will continue to be controversy over whether it is best to use one approach in preference to another but there is ample evidence in the professional literature of information retrieval to suggest that both are important for efficient retrieval of relevant records. Fidel 1992 reports that researchers in the science area (though not medicine) are more likely to use natural language or "textwords" (also called *uncontrolled terms*) than a controlled vocabulary. This also seems to be true in astronomy because until now, there has not been any standardised terminology in this field. According to Fidel 1992, "scientists tend to believe that the terminology itself is specific and well-defined". The team involved in the construction of the "Astronomy Thesaurus" (Shobbrook & Shobbrook 1993) are only too aware of the many and varied keyterms which exist to describe one concept, not to mention the use of personalised abbreviations and acronyms which have burgeoned in some new areas of research. Watson 1991 has found an astounding 7,000 unique synonyms used in astronomy.

The subject headings assigned by journals in this field are not appropriate for the on-line search procedure. They exist merely to aid in the compilation of the journal index. Until an "intelligent" interface has been designed for the searcher in the field of astronomy, it is recommended that both the librarian and the scientist use the "Astronomy Thesaurus", (Shobbrook & Shobbrook 1992). The Thesaurus was compiled for the International Astronomical Union (IAU) and endorsed by them in an attempt to standardise the terminology; it is therefore recommended for planning a search strategy. The electronic version of the IAU Thesaurus can be downloaded via ftp from the Anglo-Australian Observatory [→20]. For the methods and procedures involved in the building of the thesaurus see Shobbrook & Shobbrook (1992a, 1992b, 1993).

A thesaurus is designed for the control of synonyms, homonyms, jargon and other idiosyncrasies of the terminology. The hierarchical listing at the back of the "Astronomy Thesaurus" can assist in broadening or refining the search strategy. A multi-lingual supplement (at present being compiled in French, German, Italian and Spanish) will enable the user to interact with the database system in his/her own language. Incorporated into any computer system as a "knowledge base" for an expert system it will translate the user's request providing records/hits in all languages (Shobbrook & Shobbrook 1992).

References

[1] Adorf, H.-M., 1993: "Information-sifting front ends to databases", in: *Intelligent information retrieval: the case of astronomy and related space sciences*, A. Heck and F. Murtagh (eds.), Kluwer, Dordrecht, 49

[2] Allen, R.S., 1994: "Current awareness service for special librarians using microcomputer based Current Contents on Diskette", *Special Libraries*, 85, 35

[3] Andernach, H., Hanisch, R.J., Murtagh, F., 1994: "Network resources for astronomers". *Publ.Astron.Soc.Pacific.*, 106, 1190

[4] Borgman, C.L., Moghdam, D., Corbett, P.K., 1984: "Effective on-line searching - a basic text", Marcel Dekker, New York and Basel.

[5] Brown, M., 1994: "The role of librarians in developing software: hypertext" in: *Library without walls - plug in and go*, S.B. Ardis (comp.), Special Libraries Association, Washington DC, 131

[6] Fidel, R., 1992: "Who needs a controlled vocabulary?" *Special Libraries*, 83, 1

[7] Fisher, J., Bjorner, S., 1994: "Enabling on-line end-user searching: an expanding role for librarians", *Special Libraries*, 85, 281

[8] Fullton, J., 1993: "WAIS", in: *Intelligent information retrieval: the case of astronomy and related space sciences*, A. Heck and F. Murtagh (eds.), Kluwer, Dordrecht, 113

[9] Jackson, R. et al., 1994: "AstroWeb. A database of links to astronomy resources", *Astron. Astrophys. Suppl. Ser.*, 108, 235

[10] Kurtz, M., 1993: "Advice from the oracle: Really intelligent information retrieval" in: *Intelligent Information retrieval: the case of astronomy and related space sciences*, A. Heck and F. Murtagh (eds.), Kluwer, Dordrecht, 21

[11] Lancaster, F.W., 1972: "Vocabulary control for information retrieval", Information Resources Press, Washington DC

[12] Shobbrook, R.M. & Shobbrook, R.R., 1992a: "The IAU Thesaurus for improved on-line access to information", *Proc.ASA*, **10**, No.2, 134

[13] Shobbrook, R.M. & Shobbrook, R.R., 1992b: "The IAU Thesaurus - Towards retrieval excellence", *Astronomy from large databases II*, Haguenau, France, 1992. A. Heck and F. Murtagh (eds.), ESO, Garching, *ESO Conference and Workshop Proceedings* **43**, 479

[14] Shobbrook, R.M. & Shobbrook, R.R. (compilers), 1993: "Astronomy Thesaurus", Version 1.1, Anglo-Australian Observatory, Epping, for the International Astronomical Union

[15] Stevens-Rayburn, S., 1994: "Requirements - a librarian's point of view: a paper presented at the Workshop on Electronic Preprint Distribution Systems, 29 September 1994" (Unpublished report).

[16] Stix, G., 1994: "The speed of write", *Scientific American*, **271**, 106

[17] Van Herwijnen, E., 1993: "Intelligent information retrieval in high energy physics", in: *Intelligent Information retrieval: the case of astronomy and related space sciences*, A. Heck and F. Murtagh (eds.), Kluwer, Dordrecht, 173

[18] Warnock, A., 1994: "Project STELAR: astronomy's quiet revolution" (http://hypatia.gsfc.nasa.gov/AboutStelar/STELAR.html)

[19] Watson, J.M., 1991: "Astronomical bibliography from commercial databases" in *Databases & on-line data in astronomy*, M.A. Albrecht & D. Egret (eds.), Kluwer, Dordrecht, 199

Access Pointers

[→1] AstroWeb:
http://ecf.hq.eso.org/astroweb/yp_astro_resources.html

Library Catalogs and Services

[→2] ADS (Astrophysics Data System) Abstract Service:
http://adswww.harvard.edu/abstract_service.html

[→3] American Astronomical Society (AAS): http://blackhole.aas.org

[→4] Astronomical Observatory Trieste (OAT) library:
telnet://astrts.daut.univ.trieste.it, login: isis

[→5] Bologna Astrophysical preprints:
http://www.bo.astro.it/bap/BAPhome.html

[→6] Canada-France-Hawaii Telescope (CFHT) library:
http://www.cfht.hawaii.edu/ bryson/library.html. Preprints (of CFHT staff): http://www.cfht.hawaii.edu/ bryson/staff.html

[→7] CDS (Centre de Données astronomique de Strasbourg):
http://cdsweb.u-strasbg.fr/CDS.html

[→8] CDS Service of Abstracts (A & A, A & A Suppl. Ser.):
http://cdsweb.u-strasbg.fr/Abstract2.html or anonymous ftp:
ftp://dsarc.u-strasbg.fr/pub/abstract

[→9] CDS Service of Abstracts (PASP):
   `http://cdsweb.u-strasbg.fr/PASP.html`

[→10] Center for Astrophysics library: `telnet://hollis.harvard.edu`

[→11] Center for Astrophysics indexes to ApJ, AJ, and PASP:
   `telnet://cfa3.harvard.edu, login: APJAJ`

[→12] Center for Astrophysics ApJ Letters abstracts:
   `telnet://cf3.harvard.edu, login:APJLETT`

[→13] CERN Physics Information Service: `http://darssrv1.cern.ch/`

[→14] ESIS (European Space Information System), Bibliographic Service:
   `http://www.esrin.esa.it/htdocs/esis/esisbib.html`

[→15] European Southern Observatory (ESO) library:
   `http://http.hq.eso.org/libraries/eso-libraries.html`, library
   catalog also: `telnet://libhost.hq.eso.org, login: library`

[→16] European Southern Observatory (ESO) preprints (received from institutes
   worldwide):
   `http://arch-http.hq.eso.org/cgi-bin/wdb/eso/preprints/default`

[→17] European Southern Observatory (ESO) preprints and other in-house
   publications (electronic version): `http://http.hq.eso.org/esopub.html`

[→18] Fermilab library: `http://libmc1.fnal.gov/`, Library catalog also:
   `telnet://fnlib.fnal.gov, login: library`

[→19] Goddard Space Flight Center (GSFC) library:
   `http://www-library.gsfc.nasa.gov/`

[→20] IAU Astronomy Thesaurus: `ftp://aaoepp.aao.gov.au/Thesaurus`. For
   detailed instructions on how to download the thesaurus files contact
   Robyn Shobbrook at `LIB@aaoepp2.aao.GOV.AU`.

[→21] International Astronomy Meetings:
   `http://cadc.dao.nrc.ca/meetings/meetings.html`

[→22] Library of Congress: `http://lcweb.loc.gov/homepage/lchp.html`, LoC
   catalog also: `telnet://locis.loc.gov`

[→23] Los Alamos e-print server: `http://xxx.lanl.gov/`

[→24] NASA/IPAC Extragalactic Database (NED):
   `telnet://ned.ipac.caltech.edu, login: ned`.

[→25] National Radio Astronomy Observatory (NRAO) library:
   `http://www.cv.nrao.edu/html/library/library.html`

[→26] National Radio Astronomy Observatory (NRAO) RAP Sheets:
   `wais://annie.cv.nrao.edu/nrao-raps`

[→27] National Radio Astronomy Observatory (NRAO) preprints and published
   papers:
   `http://info.cv.nrao.edu/html/library/intro_preprints.html`

[→28] Princeton Astrophysics library: `http://astro.princeton.edu/library/`

[→29] Princeton Observatory preprints:
   `http://astro.princeton.edu/ library/prep.html`

[→30] Royal Greenwich Observatory (RGO) library:
   `telnet://gxvg.ast.cam.ac.uk, login: rgolib`

[→31] SIMBAD astronomical database:
   `http://cdsweb.u-strasbg.fr/Simbad.html`

[→32] SISSA e-print server: `http://babbage.sissa.it/` or
    `http://xxx.lanl.gov/`

[→33] Space Telescope Science Institute (STScI) library:
    `http://sesame.stsci.edu/library.html`, Library catalog also:
    `telnet://stlib.stsci.edu`, login: stlib. PASP abstracts since
    January 1993: anonymous ftp: `ftp://stsci.edu/pasp`.

[→34] Space Telescope Science Institute (STScI) STEPsheets:
    `wais://marvel.stsci.edu:210/stsci-preprint-db`

[→35] Space Telescope Science Institute (STScI) OLDSTEP:
    `wais://marvel.stsci.edu:210/stsci-old-preprint-db`

[→36] Space Telescope Science Institute (STScI) authored preprints:
    `http://www.stsci.edu/ftp/stsci/library/preprints.1st`, list of
    preprints: `gopher://stsci.edu/11/stsci/library/abstracts`, access to
    abstracts of preprints via gopher menu: .

[→37] Stanford Linear Accelerator Center (SLAC) library:
    `http://slacvm.slac.stanford.edu/FIND/books`

[→38] Stanford Linear Accelerator Center (SLAC) HEP preprint database:
    `http://slacvm.slac.stanford.edu/FIND/hep`

[→39] STELAR (STudy of Electronic Literature for Astronomical Research):
    `http://hypatia.gsfc.nasa.gov/STELAR_homepage.html`

[→40] Union List of Astronomy Serials (ULAS):
    `http://sesame.stsci.edu/lib/union.html`

[→41] UnCover: `http://carl.org/uncover/unchome.html` or
    `telnet://database.carl.org`

[→42] University of Amsterdam, Astronomical Institute (preprints):
    `http://helios.astro.uva.nl:8888/preprints/preprints.html`

[→43] University of Massachusetts departmental abstracts:
    `http://donald.phast.umass.edu/Preprints/preprints.html`

General Addresses:

[→44] Library Information Servers via WWW:
    `http://www.lib.washington.edu/ tdowling/libweb.html`

[→45] World Libraries via Telnet:
    `gopher://yaleinfo.yale.edu:7000/11/Libraries/by.place`

Newsletters that are available electronically include:

    (where an e-mail address is given, send a message to this address to be
    added to the mailing list)

[→46] AAS CSWA- Weekly women's issues / AAS Committee on the status of
    women in astronomy [CSWA]: `vaxsar.vassar.edu`

[→47] AAVSO (American Association of Variable Star Observers) Circular,
    Cambridge, Mass.: `cfa.harvard.edu`

[→48] American Astronomical Society (AAS): `http://blackhole.aas.org` or anonymous ftp: `blackhole.aas.org` includes the AAS Job Register and other AAS documents.

[→49] Bulletin sur les étoiles tardives à spectre particulier = Newsletter of chemically peculiar late-type (red giant) stars [PRG]:
(Biannual. Newsletter of the IAU working group on peculiar red giants.)
Sandra Yorka Denison University `yorka@cc.denison.edu`

[→50] Electronic journal of the Astronomical Society of the Atlantic (EJASA):
anonymous ftp: `ftp://chara.gsu.edu/pub/ejasa`.

[→51] EUVE electronic news:
Subscribe by sending a message to `majordomo@cea.berkeley.edu` and put into the body of the message "`help`" to get more information.

[→52] The HOT star newsletter:
Mostly abstracts. Philippe Eenens `eenens@tonali.inaoep.mx`

[→53] IAU (International Astronomical Union). Central Bureau for Astronomical Telegrams. Circulars / Cambridge, Mass. : Smithsonian Astrophysical Observatory:
One no longer requires a paid print subscription in order to subscribe to electronic version. Email to `iausubs@cfa.harvard.edu`.

[→54] Information bulletin on variable stars [IBVS] / Konkoly Observatory, Budapest:
Free access to current and back issues, or paid subscription. Subscribe from `IBVS@ogyalla.koonkoly.hu` or access from WWW site (`http://ogyalla.konkoly.hu`)

[→55] National Radio Astronomy Observatory [NRAO] AIPS++ quarterly report / Charlottesville, Va. : National Radio Astronomy Observatory:
Richard Simon `rsimon@nrao.edu`

[→56] The star formation newsletter:
Mostly abstracts. Bo Reipurth (`breipurt@eso.org`) or
`http://http.hq.eso.org/star-form-newsl/star-form-list.html`

[→57] ST-ECF Newsletter: `wais://ecf.hq.eso.org/stecf-newsletter.src`

[→58] STScI Newsletter:
`gopher://gopher.cic.net/11/e-serials/alphabetic/s/stsci`

[→59] [U.S. NASA NSSDC Astronomical Data Center] ADC electronic news / Greenbelt, Md.: NASA Astronomical Data Center:
`adcnews@hypatia.gsfc.nasa.gov`

Bookshops and Publishers

[→60] Addison-Wesley: `gopher://aw.com/1`

[→61] American Institute of Physics (AIP): `http://www.aip.org/`

[→62] Cambridge University Press (CUP): `http://www.cup.cam.ac.uk/`

[→63] Elsevier: `http://www.elsevier.nl/`

[→64] J.F. Lehmanns: `http://www.germany.eu.net/shop/JFL/`

[→65] McGraw-Hill: `gopher://infx.infor.com:5000/1`

[→66] MIT Press: `http://www-mitpress.mit.edu/`

[→67] O'Reilly: `gopher://ora.com/1`

[→68] SPIE: `http://www.spie.org/`

[→69] Springer-Verlag: `http://www.springer.de/`, Springer journals preview service: `http://www.springer.de/svjps/help.html`

[→70] University of Chicago Press: `gopher://press-gopher.uchicago.edu/1`

[→71] Publishers' Catalogues:
`http://herald.usask.ca/ scottp/publish.html`

[→72] Internet Book Shop: `http://www.demon.co.uk/bookshop/`

[→73] EUnet Internet Shop: `http://www.germany.eu.net/shop/`

21

How to make your information available on the network

Miguel A. Albrecht & Daniel Egret

21–1 Introduction

Scientists producing new datasets or astronomical catalogues are encouraged to make their computer–readable material and documentation available for distribution to the worldwide astronomical community. The usual procedure is by depositing a copy in one of the international astronomical data centers (see, *e.g.* CDS, chapter 16 in this volume), while simultaneously submitting a paper describing the results and procedure to a scientific journal. The journal's refereeing process will check the scientific contents while data centers normally check the technical consistency of the digital material, thus ensuring the quality of the publication.

While this is recommended in all cases, it is clear that scientists (and especially those working in a highly computerized working environment) have a lot more useful digitized information on hand than they can possibly publish by the traditional ways. The current technology allows to make this information (data, tools, news, *etc.*) available to a wider community on a fast, cheap, and efficient manner.

Different media serve different purposes. While it seems desirable to be able to read a scientific paper from a paper copy, it is uncontended that on-line systems are extremely useful for performing searches on databases, indexes, catalogs, or just any on-line collection. The advent of the World-Wide Web has opened on-line media to large scale electronic publishing. Preferred solutions seem to be:

- ▷ paper publications for scientific papers possibly including formulae and scientific rationale;
- ▷ CD–ROM's for the storage of self-contained voluminous data sets, or documents to be kept for future reference, and
- ▷ on-line publication for volatile information that needs to be distributed fast, and kept up to date.

D. Egret and M. A. Albrecht (eds.), Information & On-Line Data in Astronomy, 229–234.
© 1995 *Kluwer Academic Publishers.*

The huge growth rate of the WWW has prompted the question of quality vs quantity in the information that is made available on this medium. In fact, the medium is so easily accessible that often information is made available hastily and without the necessary considerations.

In this chapter we will advocate quality as the prime objective and at the same time encourage the utilization of the Web as an information medium. We will discuss the information handling aspects of how to organize a local information server, present a case study, and give tentative guidelines for exploiting the enormous potential of the World-Wide Web.

We will not address the technical issues of writing HTML documents and of installing HTTP servers which are extensively covered elsewhere, in books as well as on line (see *e.g.* Yahoo: HTML [→14]).

21–2 Structuring an information server

2.1 SOME EXAMPLES

Examples of current atronomical services available on the Internet have been presented by Egret and Heck (1994, [→1]). They have retained the following rough categories:

▷ astronomical institutes providing some general introductory information about the institution (see *e.g.* Lund Observatory [→3]);

▷ observatories or observing facilities providing access to documentation
  related to the instruments (see *e.g.* CFHT [→4], ESO [→5]);

▷ agencies and distributed organizations, often at an international scale
  (see *e.g.* NASA [→6], ESA [→7]);

▷ data centers and information systems providing a wide variety of possible searches under a WWW interface (see *e.g.* HEASARC [→8], CDS
  [→9]);

▷ preprint or abstract servers, and on-line electronic publications (see *e.g.*
  ADS Abstracts [→21], ADASS–III Proceedings [→10]);

▷ project-oriented networks of on-line resources (see *e.g.* WebStars [→11],
  NASA/JPL server for Comet Shoemaker-Levy 9 [→12]);

▷ yellow-page services and meta-indices of network resources (see *e.g.*
  AstroWeb [→2] and The Star*s Family of Astronomical Resources [→13]
  in the astronomical domain; or Yahoo [→15], and the WWW virtual
  library [→19] with a wider scope).

An up-to-date list of atronomical information resources can be browsed using the AstroWeb [→2] database (Jackson *et al.* , 1994). Resources found there can be used as templates when setting up your own server.

2.2   DATA AND HYPER–DATA

The first step in putting information on–line is to make a clear map of the elements (text, tables, figures, *etc.*) to publish, bearing in mind that the order in which the information will be accessed is not necessarily sequential. Hypertext makes it possible for the reader to follow a topic of interest rather than following the outline originally devised by the author: consequently a chapter may not be read from the beginning, another not read until its end. References (anchors or hyperlinks) need to be prepared carefully because they should weave a line of thought and avoid (mis)leading the reader to contexts that deal only tangentially with the topic. This is the main difference to preparing a printed paper or book. Looking at it in detail, this is not really something new: good reference books contain plenty of reference points to other publications as well as to other pages within the same volume, and therefore it is not unusual to find oneself paging back and forth. What is new in hypertext is the *availability* of information. Clicking on a link will bring the reader's attention immediately to the referred part as opposed to having to *e.g.* fetch another book from a shelf or request a volume from a library. It is in fact this availability that makes *navigation* through the information "web" possible.

In general, we can identify the following main types of hypertext references and their most common uses:

- ▷ *Table–of–contents or Portal.* Used as entry point into a collection of documents, the portal serves both as the 'piazza' (latest news, reference point) as well as the main menu display (contents, context and scope) of your information server.

- ▷ *See also.* Classical reference, pointing to other published material. This kind of link profits best from hypertext because in most cases it is possible to make a link to the actual on-line abstract of a published paper (see section 21–3) or to the text of another WWW document.

- ▷ *Name reference.* Refers to organizations or sites, projects, *etc.*, often leading to portal pages. Also used to point to further information on *e.g.* astronomical objects or specific concepts.

- ▷ *More data.* A very useful kind of links that allows the user to *e.g.* retrieve the numerical data behind a graph or have access to detailed data which would otherwise clutter the main display. Also used succesfully as a mean to organize navigation from summary to detailed views of the information.

- ▷ *Action links or buttons.* Commonly used to invoke some function within the information system. Examples include a link to mail comments to the information originator, submit a database query, *etc.*

A moderate use of links is recommended, but each database will bear exceptions to this rule. In all cases it is important to build link patterns that show consistent results when activated. For example, if you start making links in a table that point to references, this should be done throughout the table with the same rigor, making clear that when no link is present no reference is available.

It is a good policy, when including links to external resources, to mention explicitly where appropriate the name of the information provider, and to inform the *webmaster* of the referenced resource that a link has been created (so that you can be informed of any significant change on this side).

21–3   A case study

Let's look at one example of how a data collection was put on–line. We will use this example to highlight the methods and concepts used and to show how simple solutions can be quite profitable in the end.

The project itself is the catalog of Herbig–Haro Objects published electronically by Bo Reipurth [→17] and put on–line by Karen Strom, 1994 [→18].

The catalog comprises the actual table of object properties, a list of notes for each object in which observations and results are summarized and a list of references to the published literature related to the topic. Although the total number of known Herbig–Haro objects is small (a few hundred), the size of the catalog is not negligible because of the large number of publications. In fact, because this is an active field of study, a static (paper) catalog would become obsolete quite rapidly and therefore it becomes an ideal project for an on–line system.

The data was structured in a front introductory page, the object table, the bibliographical references and the object notes. The navigation nicely leads the reader from a given object to the notes which give a summary of the studies performed for that object including the observations done of that object. Each of these studies lead in term to references of the relevant papers in the literature.

The next step could be to add a link from the references to the abstract of the papers; this is one of the services offered by abstract servers like [→21] and [→9]. In fact, all what is needed is to build a *bibcode* for each reference and with that build a URL that points to the abstract (see chapters 17 and 24 in this volume).

Looking at the way the tables and files are shown, we can derive some guidelines that should help us in organizing the best possible presentation of the data.

For instance, small tables are generally to be included in the same page they are referenced to. But the solution will have to be different for long ones: many users will simply not wait until a megabyte or more has been transferred; they might simply stop the request long before the first data is displayed. A good strategy is to split the original file into suitable sizes, each of them including a "more" button at the end. Sometimes it is even possible to organize the files by, say, Right Ascension in the case of object coordinates, and include a clickable index in a "portal" page pointing to all data pages.

21–4   Some practical guidelines

The home page (or portal) of a group or institution should be fully informative for external users, about the role, location, and contact points of the organization. Because of the hypertext navigation technique, users may have direct access to almost every document within a server: it is thus a good policy to give enough practical information on each document such as signature, link to your portal, etc.

A home page should not be cluttered with too large logos and images which would make it very long to load for users connecting through a slow network. Be aware that documents will not necessarily be read through your favorite browser(s): documents should be kept browser-independent.

Some consideration should be given to the not-so-trivial issue of keeping printed and on-line versions synchronized. In many cases, the authors favourite "master" file will not be the on-line HTML pages but something like a LaTeX or ASCII document. One of the main reasons for this being the current insufficiency of Web browsers to print a document as whole —from beginning to end— including many possibly scattered pages. As a result, one finds often in the situation to have to update two versions of the same file. Some help is provided by always cleverer translators like latex2html [→16] and in many cases this will allow to keep a true master file and produce the on-line version automatically. In other cases, the solution might be to setup procedures to do the translation in the opposite direction, *i.e.* print from HTML with a fixed sequence of files.

Finally, it is important to be aware that putting information on-line is generally a specific form of *publication*: as such, the legal and ethical implications should not be neglected.

References

[1] Egret, D. & Heck, A. 1994, WWW in astronomy and related space sciences, in *Second Internat. WWW Conf.*, in press (see also: [→20])

[2] Jackson, R., Wells, D., Adorf, H.M., Egret, D., Heck, A., Koekemoer, A., Murtagh, F. (1994), *Astron. Astrophys. Suppl.* **108**, 235.

[3] Strom, K. 1994, in *Second Internat. WWW Conf.*, Chicago, Oct. 1994.

Access Pointers

[→1] WWW in Astronomy and Related Space Sciences:
http://cdsweb.u-strasbg.fr/~egret/www.html

[→2] AstroWeb: http://fits.cv.nrao.edu/www/astronomy.html

[→3] Lund Observatory: http://nastol.astro.lu.se/Html/home.html

[→4] Canada-France-Hawaii Telescope: http://www.cfht.hawaii.edu

[→5] European Southern Observatory:
http://http.hq.eso.org/eso-homepage.html

[→6] National Agency for Space and Aeronautics:
http://hypatia.gsfc.nasa.gov/NASA_homepage.html

[→7] European Space Agency: http://www.esrin.esa.it/

[→8] High Energy Astrophysics Science Archive Center:
http://heasarc.gsfc.nasa.gov/

[→9] Strasbourg Astronomical Data Center (CDS):
http://cdsweb.u-strasbg.fr/CDS.html

[→10] ADASS III Proceedings:
http://cadcwww.dao.nrc.ca/ADASS/adass_proc/adass3/cover/adass3.html

[→11] WebStars: http://guinan.gsfc.nasa.gov/WebStars.html

[→12] NASA/JPL server for Comet Shoemaker-Levy 9:
http://newproducts.jpl.nasa.gov/sl9/sl9.html

[→13] Star*s Family: http://cdsweb.u-strasbg.fr/~heck/sf.htm

[→14] Yahoo: HTML:
http://akebono.stanford.edu/yahoo/Computers/World_Wide_Web/HTML/

[→15] Yahoo: http://akebono.stanford.edu/yahoo/

[→16] LaTeX2html:
http://cbl.leeds.ac.uk/nikos/tex2html/doc/latex2html/latex2html.html

[→17] Herbig-Haro Catalog (source):
ftp://ftphost.hq.eso.org/Catalogs/Herbig-Haro

[→18] Herbig-Haro Catalog (on–line):
http://www-astro.phast.umass.edu/latex/HHcat/HHcat.html

[→19] The WWW Virtual Library: Subject Catalog:
http://info.cern.ch/hypertext/DataSources/bySubject/Overview.html

[→20] Egret and Heck 1994:
http://zaphod.ncsa.uiuc.edu/Astronomy/egret/egret.html

[→21] ASIAS (ADS Abstract service):
http://adsabs.harvard.edu/abstract_service.html

22

Computer Networking in Astronomy

Fionn Murtagh

22–1 Introduction

The academic and research community has long had a raw deal, in having to put up with a very unapproachable user interface. Other communities have demanded more. Even Unix software has been influenced, albeit with considerable delay, by the working environments of Windows and Macintosh machines.

Information systems tell a similar story. An arcane working environment slows down the user. Only users with years of experience can hope to compete (which is fundamentally unfair, since what is at issue here is not work skills, but access to information). A commercial information system like CompuServe, with over two million users, shows how information can be organized to facilitate the user. The Internet's thirty million users are now haltingly following in the same direction.

A few figures indicate current trends. Subject to continued advances (some of which will be pinpointed in this article), we believe that this momentum can be maintained in the near future. Traffic flow, in terms of numbers of bytes, on the NSF backbone showed a 114% increase between June 1993 and June 1994. World-Wide Web (WWW) usage on the NSF (National Science Foundation, USA) backbone showed a 2500% increase in the same time between June 1993 and June 1994. The number of copies of Mosaic downloaded per day from NCSA is run at 1600 during that period. Usenet newsgroups have an estimated 7,138,000 (human) readers world-wide and the number of users with accounts on machines reachable by Usenet is 22,920,000 (source: Usenet Readership Summary Report, July 1994). Frequently Asked Questions (FAQ) postings are a valuable condensation of Usenet discussions, and the number of such FAQs is currently 1964 (source: The Internet Index, 2 August 1994).

An example of extensive use of WWW and related services: "Comet explodes on Jupiter — and the Web" (K. Whitehouse, 1994). The figure 22–1 shows the file accesses on the European Southern Observatory's Web area

D. Egret and M. A. Albrecht (eds.), Information & On-Line Data in Astronomy, 235–241.
© 1995 *Kluwer Academic Publishers.*

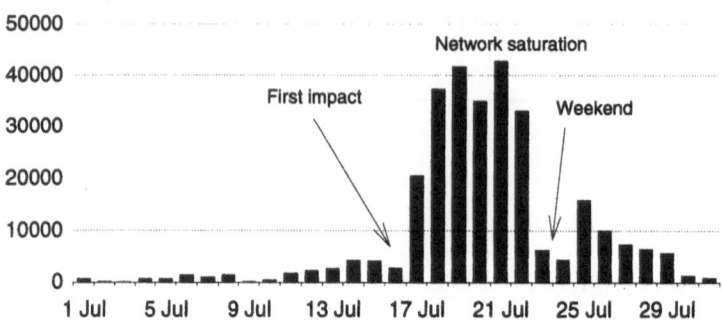

Figure 22–1: Daily accesses to the ESO WWW server during the SL-9/Jupiter collis-sion.

during the month of July 1994. The first impact in the late evening of Saturday 16 July gave rise to intense communications media interest. A marathon 28-hour press conference (which continued with periodic breaks for a week) was in session, and was backed up with information, including images, on the network. The Web bore the brunt of this, but network services which had been around for longer (email of course, and anonymous ftp, and so on) were important, as also were commercial services such as CompuServe. By the morning of Monday 18 July, ESO's Internet links had saturated and machines locally seized up with too many simultaneous users. CompuServe was already overloaded on Sunday 17 July, but the tail-off access statistics at weekends show that as of mid-1994, connectivity to the Internet was still primarily in one's office, and not at home (cf. accesses during weekend of 23/24 July). While network users in North America were certainly noticed during the comet week (the late afternoon access load on ESO's systems reflected this), it was also satisfying to note that at least two thirds of accesses were from European sites. A major site at JPL passed the 3-million mark for access a few weeks later. At around that same time a few weeks later, ESO's cumulative accesses had reached one third of a million. Certainly, better could have been done if network links were better.

The second part of the two-part mid-week peak in the graph was due to very great interest in the Q events. An email exploder served to distribute information bulletins, and a bulletin board service at the University of Mary-land was the central repository for telegram-style observing reports. Images, too, when possible, became available on the network very quickly. Within half an hour or less of an important observation, an image would be available — barely enough time to reformat and have a credit statement superimposed on it. On Wednesday morning at 3 a.m., an hour or two after some obser-vations were made at La Silla (misfortunate in having bad weather during the July events), in an almost empty ESO headquarters building — some TV crews excepted — an image hastily emailed as a rasterized sequence of integers was formatted in real time, with an impressively bright impact — on the screen and on the TV crews!

Network services had a baptism of fire during the comet week. Information distribution was possible, on a scale not experienced before. In fact, information flowed so quickly that interpretation of the speedily-unfolding events was difficult. It is interesting to reflect on the type of networking tools which, had they been available, could have helped in regard to data interpretation, and in getting accurate and penetrating insights to professionals and the interested public.

Computer-based communication has long been fundamental to the research and academic community. What is new, now, is information sharing. Even systems such as the World-Wide Web existed prior to the pioneering work by Berners-Lee at CERN, and the later development of the Mosaic browser. The real novelty in these systems and tools is client-server computing, which is part and parcel of a new distributed working environment. From information sharing to information delivery, and from information delivery to information production is a connected line of development. Note what information production encompasses: teaching course-ware, electronic publishing, and convergence with video and television, to name but a few activities.

The recent explosion of usage of client-server approaches (tools to begin with, work practices later) has lessons for policy making. Some characteristics of these approaches are (1) how they took place in a highly distributed way; (2) how European organizations like CERN played a pioneering role; (3) how seemingly diverse public and commercial bodies have been intertwined constructively in these developments.

A fuller overview of these medium-term developments will be left for *Online Data in Astronomy III*. In this article we will look over some important short-term developments. We will not deal with security and authentication. Successful and accepted standards in such areas are clearly vital for the continued survival of the Internet. We will also not deal with the current policy changes in regard to Internet funding. An interesting ball-park set of costs may be noted: according to the report [→1] authored by Henry Rzepa (rzepa@ic.ac.uk), Imperial College London, the "real" cost of networking is about 1$/MB in the US and $2–$10/MB in Europe.

22-2   WAIS and Friends

If we were writing this article one year ago, we would have begun with an overview of the WWW. With universal use of the Web, we turn instead to the issue of information access through intelligent search mechanisms.

WAIS (Wide Area Information Servers) is a widely used client-server information retrieval system. WAIS originated in a joint research project between Thinking Machines Inc. (recently experiencing financial difficulties), Apple Corp., Dow Jones & Company, and KPMG Peat Marwick, and was first released in 1991. The formation of WAIS Inc. by WAIS principal B. Kahle led to support of a freely-available version being assumed by CNIDR (Clearinghouse for Networked Information Discovery and Retrieval) , with comprehensive platform support.

Z39.50 is an information search and retrieval protocol. Z39.50-88 (Version 1) is used by WAIS. This standard is ANSI/NISO (American National Standards Institute/ National Information Standards Organization) Z39.50, which was successfully balloted in 1988. WAIS as developed by CNIDR, free-WAIS, changed its name as from March 1994 to ZDist, to signal the desired aim to move from support of Z39.50-88 to Z39.50-92.

Z39.50-92 (Version 2) is widely used in libraries and information organizations. This is an ANSI/NISO standard. This version involved alignment with the ISO (International Organization for Standardization) SR (Search and Retrieval) protocol (ISO 10162/10163), among other changes. Z39.50-88 and Z39.50-92 are not compatible. Support for Z39.50-92 will be provided by freeWAIS version 1.0 (ZDist 1.0, expected in the fall of 1994).

Z39.50 Version 3 is being balloted in the period September–November 1994 by the Z39.50 Implementors Group (ZIG), which works closely with the standard's maintenance agency, the Library of Congress. To be informed about this group's activities, subscribe to list z3950iw at address listserv@nervm.nerdc.ufl.edu.

Here are a few widely-used WAIS implementations:

▷ WAIS release 8 beta 5 minor release 1 (WAIS 8b51), released in May 1992 by Thinking Machines ([→2]). An Indiana University version (IUBio; [→3]) supported boolean search.

▷ freeWAIS versions 0.202 and 0.3, from CNIDR. Boolean search is problematic in v. 0.202. Version 0.3 supports multitype files. CNIDR can be accessed at URL [→4].

▷ freeWAIS-sf, from the University of Dortmund. This was based on freeWAIS; freeWAIS-sf(1) (or freeWAIS 1.0) was released on 9 September 1994, and further information may be obtained at [→5]. Supported features include: full boolean search; text, date and numeric field structures; configurable headlines; 8-bit character support; and other features.

Various WAIS–WWW gateways or scripts are available. For further information refer to a number of items accessible from [→6] (in turn accessible under "Manual" in the standard NCSA Mosaic help pull-down menu).

Other similar tools can also be set up as scripts. As an example, Glimpse (GLobal IMPlicit SEarch) is an indexing and querying system which allows files in possibly nested directories to be searched through. It is based on a storage-efficient index, and on `agrep` which is an extension ("approximate grep") of the well-known Unix command, which supports approximate matching (misspellings, etc.). GlimpseHTTP is an automatic HTML indexer. Glimpse is available from the University of Arizona, Department of Computer Science (for source code and documentation, refer to [→7]).

Another interesting developement relates to the Spatial WAIS standard. This is under the guidance of the Federal Geospatial Data Clearinghouse (FGDC). The aim is to incorporate a spatial metadata standard into WAIS. Although oriented towards geographic information systems (GIS), the inspirational and technical relevance for astronomy is clear. Support is available for such spatial extensions as: "at a point location"; "in a bounding box"; and "in a region". In the indexing phase, some additional index files are created to support these spatial queries. Discussion of converging such spatial prototypes into the Z39.50 Version 3 standard are conducted on list `zmap`, subscribable to at address `listserv@vinca.cnidr.org`.

22–3   Other Important Developments

We mention here two developments which are worthy of continued attention. Firstly, there is the still maturing technology of automatic network-based search and retrieval agents. Secondly, beyond the support of image, audio and video data in email – as handled by MIME (Multipurpose Internet Mail Extensions) – there is the more vital and far-reaching topic of new forms of multicast information delivery. In the latter area, we briefly discuss Internet video and audio broadcasting.

3.1   ACTIVE INFORMATION-SEEKING AGENTS

Obraczka *et al.* (1993) discuss resource-discovery approaches at a point in time just prior to the take-off of the WWW browser, Mosaic. These include X.500, archie, WHOIS, Knowbots, Netfind, Prospero (as well as WWW and WAIS).

Archie checks through registered anonymous FTP sites on a periodic basis, and arranges the file listings resulting from these automated surveys in a way which is convenient for user searches. Example: using archie, by querying on (the Internet) `resource-guide`, we found our way to `chapter.4/section4-5.ps`. We knew from a colleague's message that information was to be had there on the Knowbot service. The latter is a white pages service, embracing other services for determining email and address information.

3.2   REAL-TIME SOUND AND VISION

The Multicast BackbONE (MBONE) is a virtual network, layered on top of parts of the Internet, to support live audio and video broadcasts and multicasts. Multicast packets are encapsulated in regular IP packets ("tunneling") in order for them to be transmitted among the islands of MBONE subnets, through IP routers which themselves do not support IP multicast. The system endpoints are typically workstation-class machines. A back-of-envelope figure for traffic support is 500 kb/s. A multicast data stream may be very bandwidth-efficient since all workstations on a network can pick up the stream. Examples of MBONE broadcasts include conferences and lectures, NASA Select television, Internet Talk Radio, and regular videoconferencing. Further information is available at [→8].

Developments in relation to Internet support for multimedia are a major current spur to network infrastructure evolution.

22–4   Conclusion

The growth of the Internet is impressive. According to the newspaper *Information Systems World* (July/August 1994, p. 22), the IEEE has applied to IPng, the next generation TCP/IP (officially IPv6), for an address space for "all manufactured electronic devices". The full unleashing of portable computing is aimed at. This is all the more reason to get the house of "astronomical information" into order. Popular need for this was vividly demonstrated by the events of July 1994 when many Gigabytes of image and text data relating to the Jupiter/Shoemaker-Levy 9 encounter were downloaded per day. Openness which is an aspect of this network-based evolution in astronomy cannot but be beneficial in terms of public perception, and funding.

An example of application of searching based on text and image data was recently prototyped, using Hubble Space Telescope data. It uses textual information (proposal abstractsabstract,proposal) to query Wide Field/Planetary Camera images. It is available at address [→9]. This is based on all relevant scientific image data, and on all available proposal data. In its current implementation, a query such as "extragalactic distance scale" finds 90 ranked proposals (which contain one or more of these search terms). The text of these proposals may be viewed, and in many instances one or more accompanying inlined images are also obtained by the user. Additional linkages to the SQL-based HST image database present no problems on technical grounds. This approach to data delivery is innovative in astronomy, where – on technical grounds – image and textual data have not been merged into a seamless system in the past. Beyond such merging of particular image and text databases, there lies the great open territory of merging data in all formats. Current trends towards widespread infrastructural and protocol support of multimedia leave no doubt about the direction of these developments in the coming few years.

References

[1] K. Obraczka, P.B. Dantzig and S.-H. Li, "Internet resource discovery services", *Computer*, **26**, 1993, 8–22.
[2] (K. Whitehouse, IEEE Computer Graphics & Applications, **14**, Nov. 1994, 12–13.

Access Pointers

[→1] WWW'94 Report, Henry Rzepa:
    http://www.ch.ic.ac.uk/talks/www94_report.html
[→2] WAIS 8b51, Thinking Machines: ftp://think.com/
[→3] IUBio: ftp://ftp.bio.indiana.edu/
[→4] CNIDR: ftp://ftp.cnidr.org/pub/NIDR.tools
[→5] freeWAIS-sf:
    http://ls6-www.informatik.uni-dortmund.de/freeWAIS-sf/
[→6] WAIS-WWW gateways and scripts: http://www.ncsa.uiuc.edu/SDG/
    Software/Mosaic/Docs/D2-complex.html
[→7] Glimpse: ftp://cs.arizona.edu/glimpse/
[→8] MBONE: ftp://www.cs.ucl.ac.uk/mice/index.html.
[→9] Application for querying WF/PC images:
    http://ecf.hq.eso.org/~fmurtagh/hst-navigate.html

23

Data Storage Technology for Astronomy

Benoît Pirenne & Daniel Durand

23–1 Motivations for data storage

Nowadays, "archiving" data is a widely used expression in astronomy. A number of reasons explain it, but the more obvious ones are the cost and scarcity of large modern astronomical instruments. The existence of this book is also clue to the growing importance of this field.

Astronomy has become a "Big-Science" discipline The world largest observatories are now in the process of building huge instruments at costs which will reach astronomical heights. For instance the future European Southern Observatory's Very Large Telescope (VLT) has a planned cost of over 500 million ECU's[1]. If one expects the facility to be operational 30 years long and to be used 300 nights per year, that comes to a nightly cost of over 50,000 ECU per night, to which one has to add the running costs. This is a very expensive gift to any successful program. It is clear that the output of any particular scientific night should be shared. Hence the data must be archived to made available to the largest community of scientists.

It should be mentioned that the VLT is not the only big science venture on the ground: the Japanese "Subaru", the US "Keck" and the international "Gemini" projects share all the same characteristics.

Many of other arguments in favor of data archives exist:

▷ The high demand for top-of-the-line telescope-instruments-detectors setups. Due to their rare characteristics, many scientists will try to get access to them and an over-subscription rate of 5 is already common for such resources. Another way to make the instrument performance accessible to more users is to build an easily accessible archive.

▷ The present day recession and financial difficulties of countries. Science being one of the first casualties in economic bad times, it should be

---

[1]Throughout this paper we will chose ECU as currency to describe all prices. The reason being that the the technology survey was done in Europe, using local vendors' quotes. At the time of this writing 1 ECU $\simeq$ US$1.25

D. Egret and M. A. Albrecht (eds.), Information & On-Line Data in Astronomy, 243–257.
© 1995 *Kluwer Academic Publishers.*

clear to most of us that the current large instruments might be the last ones before long and that some of the ones being now designed and constructed will be de-scoped or delayed (indefinitely?). For the next generation astronomers, now teens dreaming of the stars, we have to keep as much as possible of what the current and previous generations of scientists could achieve with taxpayers' money.

▷ Scientists having successfully applied for time on an instrument are going to look at the targets chosen with a certain purpose in mind. This purpose might overlook some other potential uses of the data. Therefore, the data must be preserved for other scientists who might have other ideas as to the use of it.

▷ New scientific methods are also arising from the existence of archives: multi-wavelength analysis, object-type oriented studies, automatic searches for object morphologies, statistical studies, etc. are now becoming possible thanks to the existence of those archives and the availability of cheap "CPU cycles".

Space missions and archives: the precursors   The first groups to make use of digital archive in astronomy were to be found in the space community. Because of the uniqueness of the instruments, their costs and the fact that the data could only be in electronic form, the reasoning described above took place. Successful current examples are IUE (Wamsteker, 1991) and HST (Pirenne *et al.*, 1993; Pasian *et al.*, 1993). Not to be overlooked is also IRAS with the record-breaking number of papers per operational years. Ground-based observatories started some archival ventures in the early eighties but they do not seem to be so well known (Raimond, 1991). Radio astronomers were also early players in this game (Wells, 1993).

A Review of Problems and Solutions   Since many observatories are now considering digital archives as a way to further the lifetime and multiply the production rate of their most valuable instruments, the time was ripe for a review of the technology available in this field but also, and more importantly, to insist on the major problems of archives: the chasm between the expected lifetime of any storage technology and that of the data. Solutions to overcome this problem will be described. It will be shown that only sound and evolution-oriented design phase will make for a successful archive.

23–2   Definitions

Before to proceed with technicalities, it is important to define a number of points. Among them: *Storage Technology* —the subject of this paper— but also the *Lifetime of storage technologies*; the *Dynamic and Static capacities* of the archive system.

Storage Technology The storage techniques used in modern computer systems can be decomposed into a set of several different hierarchical levels.

Level	Description	Example	Category
1	storage *medium*	CD-ROM, tape reel, ...	Hardware
2	*reading device*	disk drive, tape drive	Hardware
3	data structure on medium	file system	Software
4	control & management procedures	archive storage and re-distribution	Software

Table 23–1: The various hierarchical levels of storage technology

These levels (see table 23–1) are all necessary for an archive to work. Level 4 is explained in some details by Pasian and Pirenne (1992) for the case of the HST/ESO/CFHT archive. Level 3 should be covered by industry or international standards. But this is not always true: For instance optical disks do not enjoy an internationally recognized standard. Therefore those who want to use this technology must usually define their own [11] or buy a commercial proprietary system. Levels 2 and 1 usually are a question of choice. We will cover those levels in the next sections.

Lifetime of Storage Technologies Here we touch on one of the crucial problems of archives. Repositories of information are supposed to maintain it safe and legible over long periods of time (Pirenne *et al.* , 1993). That is, both hardware and software should live long. However, if the software can usually be adapted with some effort to cope with the steady computer — and user needs— evolution, the hardware situation is more of a problem. Both the storage medium and —even more— its reading device decay. In some cases, the medium lifetime can be longer than its reading device's. For example, one of the reason why the HST archive has moved from one storage system to a more modern one is because the vendor of the optical disk readers and platters used so far would not support any of the two and becomes increasingly reluctant to repair the drives or to manufacture media. The existing media already written will probably still be readable in 20 years, but from now on it becomes increasingly expensive to maintain a reading capability! Also, choosing a higher density medium is usually a cost-effective measure. Migrating to a more modern optical disk technology will allow the HST archive to divide its media costs by 3 for all future data.

The archives from a static and dynamic standpoint The archive systems must be designed with sufficient capacity to store all the data produced by the observatory throughout its lifetime that is, science data and calibration data. The archive must be designed with growth in the plan. It is not sufficient to plan the system with just enough capacity for a current observatory set up: instruments evolve and produce more and more data.

Moreover the archive system must be designed to sustain the data input/output rate: in order to avoid backlog in archiving and long wait queues in the data distribution service, it is necessary to design data channels and recording/retrieving performance that will manage to cope with the throughput and its rate evolution.

23–3   Technological Choices

Designing an archive system implies making a number of technological choices which will only be valid for a certain time. These technological choices can be "modeled" as a function of 3 parameters, each function of time. One could write the following "formula":

$$Techno\_scheme = F(avail\_tech(t), input(t), output(t))$$

Where *input*, the input data volume is normally a constant in the case of a space-borne mission and evolves as a staircase function for lively ground-based observatories to follow the trend in detectors (see *e.g.* Pirenne, 1994).

The *output* parameter (re-distributed data volume per unit of time) has a slightly different shape, which the experience shows as model-able using a logarithmic function. For very successful observatory/space mission, the data output rate can sometimes outgrow the data ingest rate. The example of IUE is very useful in this respect (Wamsteker, 1991).

The *available_technology* parameter will be discussed below as far as its present state is concerned. In the past years, fortunately, its evolution has followed the trend in detectors and it has been possible for most archives to be set up as off-line data centers at reasonable costs for most lively ground-based observatories or space missions. Completed mission archives can, usually within a few years' time, be migrated from off-line archives to on-line or easily reproduce-able high-density media when the technology catches up.

3.1   THE HARDWARE

Within the limits of the available technology, and with some reasonable model of the data volume, input and output rates, the choice of tools for running an archive will also be guided by the following —sometimes contradictory— constraints:

▷ Storage: is best done in a sequential (chronological) manner and is best served by cheap sequential access devices (e.g., tapes).

▷ Storage: must be done at the end of an analysis and possibly also after a calibration pipeline. Data must undergo a minimum of check and standardization before to be permanently stored.

▷ Storage must not slow down retrieval operations.

▷ Storage should be done in a software format that will survive as long as possible and is a standard: e.g., data in a computer/software/medium independent format (FITS).

▷ Distribution must serve requests in a predictable and timely fashion.

▷ Distribution must provide the choice of various standard output media (for off-line data) and should use the standard FITS format.

▷ Distribution deals with requests for files not usually ordered in a chronological way. Therefore, best data re-distribution system will use random access storage systems.

The current choice in techniques suitable for archiving can be structured in the way presented in table 23–2. It is important to note that the computer market is now more and more offering Juke-boxes, also named auto-changers. They are available for most of the existing removal media listed in table 23–2. They limit operator intervention and in the case of "RND"-type media, allow for quasi on-line access times. The characteristics of some of those types are analysed and compared in table 23–3. The list is not exhaustive, it does not for instance consider the data "vaults" and other multi-terabyte data "silos". This study takes into account systems allowing 1-2 terabyte in maximum configuration only. This amount represents the needs of most astronomical archives so far. But it is clear that those figures will have to be revised in the future.

Notes on table 23–3

▷ Prices based on upper limits found on the European market in the Spring on 1994. (Prices without local tax). Please note that this list will be kept up to date and will be made available for public consultation on Internet using the access URL [1].

▷ RND-OPT-CDx (CD-ROM) are becoming an interesting alternative to RND-OPT-WRM (Write once read many optical disks). It is now possible to get cheap writers and fast readers. Even though the individual volume capacity is modest compared with the larger optical disks, the availability of large juke-boxes makes it a cheap and convenient prime storage system. Moreover, CD-ROM's enjoy an international write format, shared by many vendors, therefore ensuring a fairly long system lifetime. New standards for this medium are also emerging: a number of companies are currently developing prototype recorders with respectively 3 and 10 times the present density. The former would simply increase the density on the medium, whereas the latter would be based on a multi-layer recording technique. This will be a tremendous improvement with respect to present-day standards and might install CD-ROM as the media of choice for quite some time.

▷ SEQ-MAG-HSC devices (DAT's and Exabytes) currently enjoy a great popularity due to their high density and the very low price per GB. The standard file seek time is however very bad on these devices and the figures presented in table 2 can only be reached if a special software driver is installed in the computer operating system. Standard file random access times are normally of the order of an hour!

▷ SEQ-MAG-HSC devices (DAT's and Exabytes) can be purchased with internal hardware data compression. However, for archival purposes, the driving computer can also compress software-wise. For archive products distribution, compression is *not* recommended. Moreover, the data compression algorithms used on the hardware are usually best suited for ASCII data. In particular astronomical images will *not* compress very well using those techniques. See also the section on data compression.

▷ SEQ-MAG-HSC devices (DAT's and Exabytes) when used for data distribution should only use the basic common denominator model that most users can be expected to read easily: 2.5GB first generation Exabyte; standard DAT/DDS cassettes. The figures in table 23–3 describe however the uncompressed, higher density models (Exabyte 85xx and DAT/DDS-2 with 120m tapes respectively).

▷ The real capacity of SEQ-MAG-HSC devices (DAT's and Exabytes) can be very different from what is announced by vendors: the inter-file gaps on the tape occupy a non-negligible space, which is a problem for tape written with the FITS standard. A simple test could write 700 4.25 MB FITS files on an EXABYTE 85xx (roughly 3 GB/5 GB). The DAT did obviously better with 1.9 GB/2.0 GB. Please note however, that the newest Exabyte models can write smaller inter-file gaps. But one can question backward compatibility.

▷ The lifetime of the various devices are values provided by vendors for most devices except for the SEQ-MAG-* categories where there is sufficiently long experience for a personal assessment. The question marks on the longer lifetimes insist on the fact that no practical experience can possibly prove or disprove those figures.

▷ SEQ-MAG-HSC-ID1 is the very recently introduced digital data cartridge system. It has the advantage that it relies on the ANSI ID1 standard, which is a promise of multi-vendor availability.

3.2 DATA RE-DISTRIBUTION USING COMPUTER NETWORKS

The advent of better and faster network connections between archive centers and end-users has changed a lot the data distribution process. For most computer users, tools like ftp (possibly via the popular NCSA Mosaic interface) have become a convenient way to access remote public data and software. Distributing archived data using the "net" has substantial advantages for both end users and archive centers: faster, reliable, less manpower needed to prepare distribution media, etc... This distribution technique will probably become more important than any other "hard" distribution medium in the near future. Tapes/CD-ROM distribution will remain necessary only for large quantities and/or systematic data distribution tasks. For example, distributing hundreds of mega-bytes of Principal Investigator's data on a daily basis should still be done using tapes.

We have added here information on getting network connection, but the costs vary substantially from country to country, even within Europe, and there is almost an order of magnitude of difference for the price of the same services between North America and Europe. In this context the following lines have to be taken with extreme caution and should be used as a point of comparison only. The amount of options available for connections makes it also difficult to suggest a universal solution.

Basically, two types of end-users connections to the international networks (Internet) can be considered: a permanent, leased-line connection (a must for professional institutions) or a temporary, few hours/day open public telephone connection. This is more appropriate for individuals.

The price in each category has also two parts: the cost of the raw connection and the price of the Internet service provider.

Table 23-4 gives a few examples for different network bandwidths. Private temporary connections are not considered in this survey. Two options are however available to individuals: the standard telephone line with a maximum authorized of 28800 Bauds (limit frequently revised) transmission rate or an ISDN (Integrated Services Data Network) connection with 2x64KBauds.

3.3 DATA COMPRESSION

Data compression has become fashionable in the past three years. Two types of compressions are available: destructive (lossy) and non-destructive (non-lossy). Each has its own application field: lossy compression will usually be applied to images (ie. more than 1-D datasets) when quick data transfer is required and the resolution of the original image can be damaged without harming the application/visual quality. Destructive compression can reduce the size of an astronomical image by a factor of 200 (see Hill *et al.* , 1994) without losing too much the semantical content. Non-destructive compression is used for data transfer of other type of datasets and distribution and obviously compresses not so well as the previous method: a factor of 3 in size is usually the best one can get nowadays for astronomical data. It has however to be borne in mind that this field is still under development and that better methods may still appear.

Archives can not usually forecast the future use of their data. That means lossy compression is to be rejected from the design as a possible way of saving prime storage space. Non-lossy compression or no compression at all is now a trade-off between the time spend in compressing/decompressing and the amount of space gained on the medium. In other words, this discussion comes down to comparing the cost of the CPU with the cost of the media, convolved with the time constraints in retrieving the data, and how many times one single dataset will be retrieved.

Data compression is available hardware- and software-wise. For astronomical purposes, where images are usually noisy, the existing hardware compression systems are usually not well suited. They are normally designed for TV broadcasts (in a lossy form *e.g.* MPEG) or for backup purposes (non-lossy compression). The latter best compresses text data.

For image storage purposes, it is better not to rely on vendor–provided techniques but rather on published image compression algorithms such as FITSPRESS (Press, 1992 and URL [→2]) or one based on the H-transform (*e.g.* White, 1992 and URL [→3]). Another method (COMPFITS, [→4]) is also available in Veran and Wright (1994). Table 23–5 gives a non-exhaustive list of some of the widely available data compression systems.

3.4 JUKE BOXES (AUTO-CHANGERS)

The role of juke boxes is to place a large collection of off-line data volume in special cupboards in order to allow for a rapid insertion in a reading device using a robotic arm. The data made available in a juke box or auto-changer is often qualified as "semi on-line". When dealing with random access devices (*i.e.* not tapes), the file access time is only handicapped by the volume change operation, which takes a few seconds. This technique is very useful for unattended random retrieval of files from mid-size to large archives. However, the technique will really pay off if the amount of data to be made available for rapid retrieval does not exceed the capacity of 2 to 3 such juke boxes, otherwise the costs will become unbearable. Also, one should keep in mind that juke boxes will not remove entirely the manual operator work: from time to time new media must be entered in or removed from the juke box. If retrieval time is not an overwhelming constraint, and if the archive cannot be held semi on-line with a reasonable amount of juke box units, human operators should be preferred. The example of the European copy of the HST archive is such a case (Pirenne, 1993).

As far as the capacities are concerned, the largest optical disk juke boxes can hold over 1000 GBytes, and the largest Exabyte or DAT auto-changers will hold up to 580 GBytes. (All figures with uncompressed data and standard drives).

It is worth noting the existence of more and more CD-ROM juke boxes with capacities ranging from a handful of disks to about 500 volumes. With the promised sky surveys widely distributed on several dozens of volumes, these devices will find a role in the larger institutions where the demand for this type of data will justify it.

A recent improvement is also the fact that affordable CD-ROM writers can now be inserted in those jukeboxes. CD-ROMs can now be seriously considered as the prime storage support: on top of the advantages mentioned before, CD-ROMs can now be conveniently handled in an automatic way that overcomes their relatively low capacity.

23–4   Databases

An astronomical catalogue is generally composed of one or several tables with rows and columns of different types. This structure is best dealt with with relational database management systems (RDBMS). A brief description of RDBMS's is given, *e.g.* , by Pirenne and Ochsenbein (1991).

Not only do RDBMS's allow easy manipulation of our catalogues, but they also offer a number of other advantages that can be put at use to serve other aspect of archive operations. To summarize, here is a list of features provided by RDBMS's and their potential use in archive systems.

   ▷ Handling of data structures that can be represented in rows and columns.
   ▷ Support of most data types of use in use by scientific applications (integers, floating point numbers, strings of characters).
   ▷ Support for extra optional data types (date/time, text, image, ...)
   ▷ Universal query language (SQL) in front of which more friendly user interfaces can be built.
   ▷ Possibility to merge, join, correlate different tables.
   ▷ Security (backups & recovery well integrated).
   ▷ High access performance.
   ▷ Client-server design allowing simultaneous data access by many different users at geographically distinct locations.

▷ transaction systems which control simultaneous writes on the same table.

▷ can be the central control engine for all archive operations.

Disadvantages related to the use of RDBMS's for scientific archives could be:

▷ Not all the necessary data types are available (multi-dimensional coordinates system cannot be represented as one entity),

▷ Transaction oriented systems are good for commercial applications (lots of queries, small amount of results): need for tuning the system to scientific type of queries (fewer queries, lots of results).

▷ Free text search not well supported.

▷ Commercial systems are fairly expensive to purchase and maintain.

▷ Need a knowledgeable person to administrate the system.

▷ If the RDBMS is used to control the entire archive (data access, request service management, housekeeping etc.), the RDBMS quickly becomes central to the archive and hence a single point failure.

The purchase of an RDBMS for supporting the archive must be accompanied by a commitment for support. It will facilitate a lot of the tasks, will take care of many problems, will allow nice features to be developed, will give some kind of security feeling, but there is a price to pay for it: a purchase price, a maintenance price and a manpower price.

The purchase and maintenance price problem can somewhat be alleviated by the choice of a public domain system but no guarantees for support can be offered. Such a choice will imply an even more knowledgeable person to take care of the system. A list of those free database systems is available at [→5].

23–5 Conclusions

When setting up archives, a number of questions arise. Besides those dealing with the presently available technology for solving present time problems, these questions usually deal with the long term prospects of the chosen technology. We have seen in the previous section the current possibilities. Let us now say a few words about the future.

The astronomical data production rate is nowadays increasing at a tremendous rate: the average size of CCD's images has doubled every other year and the prospects for the near and medium term future call for a continuation of the current trend. There are many examples of CCD mosaic being planned, built or even in use to demonstrate the archival challenges for the years to come. Will the storage devices support this exponential data growth? The answer is probably yes, since the last ten years of computer history have showed a similar —if not more sustained— trend towards larger volume capacities.

Another related problem is the data transfer from the CCD camera to the archive center and from the archive center to the final users. The developments in computer networks have shown large increases in both bandwidth and reliability. The growth of the data production rate of future astronomical detectors following a parallel trend is also a forecast that one can venture without too many risks.

The processing of larger and larger amounts of data are of course also a concern for archive centers which sometimes are also given the responsibility of calibration pipelines. Again one need not be overly optimistic to confidently admit that the processing capabilities of tomorrow's computers will continue to overpower the data amount growth. To illustrate this assertion: it is today far less expensive to purchase a computer for astronomical data handling of say, 4K by 4K CCD's than it was 7 years ago for 512 by 512 pixels.

There is nowadays a fair amount of astronomical archive centers with a lot of experience to share with would-be archives. Today, it is no longer necessary to invest years of efforts in re-inventing software and hardware solutions when setting up a project. Looking around and adopting proven solutions will allow one to concentrate on the really important aspect of archives: the data, rather than on the tools to manipulate them.

References

[1] Hill, N., Crabtree, D., Gaudet, S., Durand D., Pirenne, B., Irwin, A. (1994), *Generation and Display of Online Preview Data for Astronomy Data Archives*, PASP Conference Series, Astronomical Society of the Pacific, **61**, Astronomical Data Analysis Software and Systems III, D. R. Crabtree, R. J. Hanisch, and J. Barnes, eds., 115.

[2] Pasian, F., Pirenne, B. (1992), *Migration to Unix of the ST-ECF Archive*, ST–ECF Newsletter **17**, 20.

[3] Pasian, F., Pirenne, B., Albrecht, R., Russo, G. (1993), *The European HST Science Data Archive*, Experimental Astronomy **2**, 377.

[4] Pirenne, B. (1993), *Two years of HST Archive operation – first conclusions*, ST–ECF Newsletter **19**, pp. 18-19

[5] Pirenne, B. (1994), *Choosing the Right Tools for an Archive* Workshop on Handling and Archiving data from ground-based observatories, Trieste, Italy, May, 1993, .

[6] Pirenne, B., Ochsenbein, F. (1991), The Guide Star Catalogue in STARCAT, ST–ECF Newsletter, **16**

[7] Pirenne, B., Benvenuti, P., Albrecht, R., Rasmussen, B.F. (1993), *Lessons learned in setting up and running the European copy of the HST archive*, SPIE International

Symposium on Aerospace and Remote Sensing, Space Telescopes, Orlando, FL, April 1993, Proceedings in Press.

[8] Press, W. H. (1992), *Wavelet-based Compression Software for FITS images*, Astronomical Data Analysis and Systems, Proceedings, Astronomical Society of the Pacific Conference Series, **25**.

[9] Raimond, E. (1991), *Two Observatories and their Archives: the Isaac Newton Group of telescopes, La Palma and Westerbork Synthesis Radio Telescope*, Databases & On-Line Data in Astronomy, M. Albrecht and D. Egret eds., Kluwer Academic Publishers, 115.

[10] Raimond, E. (1994), Workshop on Handling and Archiving data from ground-based observatories, Trieste, Italy, May, 1993, .

[11] Russo, G., Russo, S., Pirenne, B. (1993), *An Operating System Independent WORM File System*, submitted to Software - Practice and Experience.

[12] Taylor, B. G. (1992), *Space Astronomy from Archives and Data-Bases: the European Context*, Astronomy from Large Databases II, Haguenau, France, September 1992, Heck, A. and Murtagh, F. eds.

[13] Veran, J. P., Wright, J. R. (1994), *Compression Software for Astronomical Images* PASP Conference Series, Astronomical Society of the Pacific, **61**, Astronomical Data Analysis Software and Systems III, D. R. Crabtree, R. J. Hanisch, and J. Barnes, eds., 519.

[14] Wamsteker, W. (1991), *The many faces of the Archive of the International Ultraviolet Explorer satellite*, Databases & On-Line Data in Astronomy, M. Albrecht and D. Egret eds., Kluwer Academic Publishers, 35.

[15] Wells (1994), D., , Workshop on Handling and Archiving data from ground-based observatories, Trieste, Italy, May, 1993,

[16] White, R. L. (1992), *High Performance Compression of Astronomical Images*, available from `ftp://ftp.stsci.edu/`.

Access Pointers

[→1] Data storage devices:
    `http://arch-http.hq.eso.org/Data-Storage-Technology.html`

[→2] FITSPRESS: `ftp://128.103.40.79/pub/fitspress08.tar.Z`

[→3] H-compress: `ftp://stsci.edu/software/hcompress/hcompress.tar.Z`

[→4] COMPFITS:
    `ftp://ftp.cfht.hawaii.edu/pub/compfits/CompFITS-09Mar92.tar.Z`

[→5] Free database systems:
    `ftp://idiom.berkeley.ca.us/pub/free-databases`

TYPE	Description	Examples	Recommended Usage
RND-IL	Random Access Silicon-based "disk"	RAM-disk	very short term storage
RND-MAG	Random Access Magnetic disk	Standard mag disk	Data processing, staging space
RND-OPT-CD	Random Access Optical CD-ROM	CD-ROM device	wide distribution of small to mid-size archives
RND-OPT-WRM	Random Access Write-once/Read-many Optical disk	5 1/4" or 12" optical disk	Larger archives
RND-MOP	Random Access Magneto-Optical disk	5 1/4" re-writeable optical disk	Short time archiving of small to mid-size archives
SEQ-MAG-HSC	Sequential Access tape, Helical scan	4-mm DAT/DDS or 8-mm Exabyte	backup, data distribution
SEQ-MAG-STK	Sequential Access tape, Serpentine track	QIC-xxx tapes	backups, software distribution
SEQ-MAG-PTK	Sequential Access tape, parrallel tracks	9-track reel tape	old-fashioned?
SEQ-OPT	Sequential Access optical tape	1-TByte optical tape	large archives with limited file retrieval

Table 23-2: Structure of the various types of storage devices commonly available today

TYPE	Drive Cost ECU	Media Cost BCU/GB	Rnd File Access sec.	Transfer Rate MB/s	Vol Cap. GB	Medium Lifetime year	Description
RND-SIL	N/A	68000	0.0007	100	0.5	N/A	RAM-disk
RND-MAG	4500	500	0.01	10	9.0	10	Standard mag disk
RND-OPT-CDW	5000	12	0.2	0.2	0.6	>30?	Writeable CD-ROM
RND-OPT-CDR	500	N/A	0.2	0.4	0.6	>30?	Mass-produced CD-ROM
RND-OPT-WRM	20000	65	0.6	0.3	10.0	100	12" WORM opt. disk
RND-MOP	3700	200	0.2	<2.0	1.2	>10?	Re-writeable OD
SEQ-MAG-HSC-DAT	1250	5	<60	0.5	4.0	>3?	4-mm DAT/DDS-2
SEQ-MAG-HSC-EXA	2500	2	<60	0.5	5.0	>2?	8-mm Exabyte
SEQ-MAG-HSC-ID1	250000	?	?	>10.0	100.0	?	Digital Data Tape
SEQ-MAG-STK	1200	20	?	?	2.5	5?	1/4-inch QIC-2GB
SEQ-MAG-PTK	7000	120	60	0.6	0.2	5	9-track reel tapes
SEQ-OPT	200000	5	<60	1.5?	1000.0	>20?	Digital Optical Tape

Table 23–3: Review of the individual data storage devices available as of Spring 1994 and their physical and economical characteristics (see text)

Bandwidth costs in Europe		
Bandwidth	Raw leased-line cost	Internet connection cost
64Kbps	$\simeq 500/month$	$\simeq 2500/month$
128Kbps	$\simeq 900/month$	$\simeq 4500/month$
256Kbps	$\simeq 1800/month$	$\simeq 8750/month$
2Mbps	$\simeq 15000/month$	?
Bandwidth costs in the U.S. (Los Angeles)		
1Mbps	$\simeq 650/month$	?s

Table 23–4: Examples of data network costs. Figures are highly volatile and vary a lot from country to country. They should not be used verbatim but as rough estimates. They are based on an enquiry in Western Europe in late 1994. Prices in ECU, for a point-to-point connection within the same telephone area. The U.S. (extreme) example serves as a comparison to demonstrate the yawning gap between the two continents.

Name	Lossy?	H/W–S/W?	Application field	Comment
Unix Compress	N	S/W	ASCII/Binarys	standard, not very efficient
GNU Zip	N	S/W	ASCII/Binary	newer, better
FITSPRESS	Y/N	S/W	FITS files	one of the first
H-transform	Y/N	S/W	FITS files	really tuned to astronomical images
JPEG	Y	S/W	images	gen. purpose image representation system (int. standard)
MPEG	Y	S/W–H/W	movies	int. standard for movies
IDRC-x	N	H/W	gen. purpose	Exabyte compression system
DDS-DC	N	H/W	gen. purpose	DAT compression system

Table 23–5: Review of some of the most well known image compression systems. This list is non-exhaustive.

24

NED and SIMBAD Conventions for Bibliographic Reference Coding

M. Schmitz, G. Helou, P. Dubois, C. LaGue, B. Madore,
H. G. Corwin Jr. & S. Lesteven

24–1 History

The uniform 19–digit code used for bibliographic references within NED and SIMBAD was developed by both teams in consultation with Dr. H. Abt, editor of the *Astrophysical Journal*. The primary purpose of the "REF_CODE" is to provide a unique and traceable representation of a bibliographic reference within the structure of each database. However, in many cases, the code has sufficient information to be quickly deciphered by eye, and it is used frequently in the interfaces as a succinct abbreviation of a full bibliographic reference. Since its inception, it has become a standard code not only for NED and SIMBAD, but — with minor variations — for ADS and other bibliographic services. In addition, the acronyms for journals used as part of the code have become standards for some of the main astronomical journals in their own bibliographies.

Our main consideration in designing the REF_CODE was to make its definition as objective as possible. This helps to avoid having the history of data entry affect the naming system; allows automatic coding to some extent; avoids confusion, conflicts, and ambiguities in its meaning; lets different individuals or teams construct REF_CODEs without having to resort to constant consultation on the details of the code; and facilitates exchange between databases (*e.g.* NED and SIMBAD).

24–2 Definition

The standard code is a string 19 characters long, a combination of fields, some numerical and some alphabetic, exactly predictable for journal articles, but not necessarily for books. The format is as follows, with the various fields

D. Egret and M. A. Albrecht (eds.), Information & On-Line Data in Astronomy, 259–270.
© 1995 *Kluwer Academic Publishers.*

explained below. Blank spaces within the string are replaced with periods, and no leading zeros are allowed in volume and page numbers.

YYYYJJJJJVVVVMPPPPA

YYYY: The four digits of the year of publication.

JJJJJ: Code for the publication, entered left–justified within the five spaces. Five categories are distinguished:

PERIODICALS (including both regularly–published periodicals and occasional publications): these codes are acronyms based on the names (as in ApJ, A&A, PASJ, MNRAS), and are reserved for all years. The codes for the journals that NED presently scans directly are given in Table 24–1. Codes for journals currently scanned for the SIMBAD bibliography are given in Table 24–2, and a sample of codes for less–frequently encountered journals are given in Table 24–3. A complete listing of these tables is available on the World-Wide Web at [→1] and [→2].

CATALOGS: these codes are generally built from "standard" abbreviations of the catalogs' names. Examples are UGC, ESO, RSA, and RC3. If the catalog is a multi–volume work, the volume number is inserted in the Volume field (see below). The codes for some often–used extragalactic catalogs are listed in Table 24–4.

BOOKS (by which we mean all other monograph–length publications): the codes in this category are constructed in essentially the same way as those for periodicals and catalogs, from some or all of the initials (or following letters) of the title. While there is clearly some freedom in assigning codes to books, it is not necessary for the user to be able to identify a random book from its reference code (the database interface does the decoding as needed). Note also that the same code combined with a different year points to a different book.

THESES (primarily doctoral theses, but occasionally includes masters theses): these codes are acronyms based on the name of the university granting the degree (see Table 24–5 for examples; the complete list is available on–line). For theses, the volume number field ("VVVV" below) contains ".T00". In the case of duplicate author initials, the ".T00" becomes ".T01", ".T02", *etc.*

UNPUBLISHED: this, unfortunately, is unavoidable as a category. If the reference is to a collection of data never described in print, then this field will contain the code "UNPUB". Private communications to NED or SIMBAD carry the code "PrivC".

VVVV: Volume number, right–justified, if the reference is to a periodical; otherwise, the second character in this field is a letter that serves as a classification flag. The following flags and classes of books are presently identified:

B	textbook
C	catalog
M	digitized version (magnetic tape, CD-ROM, *etc.*)
P	preprint
R	report or conference proceeding
S	symposium
T	thesis
U	unpublished

For multi-volume books, catalogs, and reports, the volume number is given in the last two digits.

M: This field is intended to break any remaining ambiguity after volume number, page number, and author's initial have been specified. It is used only when necessary, as in the following two classes of problems:

One class of ambiguities results when there are two or more independent page sequences within the same volume number, in which case the following codes are reserved for this field:

L	Letters sections in various journals.
p	Pink pages in *MNRAS*.
a, b, ..., z	Issue numbers within the same volume, each of which starts with page 1 (*e.g. Physics Today*).
A, B, ..., K	Issue designations used by publisher within same volume, where each issue starts with page 1.

Another class of ambiguities results when there are two or more articles on the same page, as in *Nature*. Such articles starting on the same page are numbered sequentially in their order of appearance, and a code corresponding to this order is inserted in this field. In that case, the code has values

Q, R, ..., Z   First, second, ..., tenth article on the page.

For Theses, this field contains the author's first initial.

PPPP: Page number of reference, or "...0" when the whole book is referenced. This field contains the page numbers, which are right–justified within the four spaces available, preceded by periods to fill empty spaces.

A: This field contains the first letter of the first author's last name. This provides some redundancy in the code which might be useful in tracking down errors. If the first author cannot be identified, or no authorship is expressed, a colon (:) appears in this field. When the REF_CODE as a whole does not follow the standard rules described above (which might happen for books) a percent sign (%) is inserted in this field. This field is case sensitive.

Here are some examples to illustrate the use of the reference code:

`1983ARA&A..21..177S`	Stein and Soifer. 1983, *Ann. Rev. Astron. Astrophys.*, **21**, 177.
`1988ApJ...324..767W`	Ward *et al.* 1988, *Astrophys. J.*, **324**, 767.
`1988ApJS...66..183J`	Jura. 1988, *Astrophys. J. Suppl.*, **66**, 183.
`1988PASP..100..625S`	Sandage. 1988, *Publ. Astron. Soc. Pacific*, **100**, 625.
`1988Natur.331.6157B`	Bergvall. 1988, *Nature*, **331**, 6157.
`1976ApJS...31..187D`	Dressel and Condon. 1976, *Astrophys. J. Suppl.*, **31**, 187.
`1978IAUC.3305....1K`	Kowal, Lo, and Sargent. 1978, *IAU Circ.* No. **3305**.
`1988A&A...206L..23M`	Maurogordato *et al.* 1988, *Astron. Astrophys.*, **206**, L23.
`1984IRSD..R....118G`	Gatley. 1984, in *Lab. and Obs. IR Spectra of IS Dust*, proc. of the Hilo Workshop, July 1983, ed. Wolstencroft and Greenberg, p. 118.
`1909UCB...T00E....F`	Fath, E. A. 1909, *The Spectra of Some Spiral Nebulae and Globular Star Clusters*, thesis, Univ. of Calif., Berkeley.

24–3   Conclusions

The Bibliographic Reference Code is a domain–specific code which was designed to be sufficient for the immediate needs of astronomy in uniquely, succinctly, and informatively identifying bibliographic references. Nevertheless, the REF_CODE proved to be general enough to encompass most of the existing astronomical literature. But these REF_CODEs were not explicitly designed to be so general that they were guaranteed to automatically encompass all presently available media, nor do they necessarily fully anticipate future directions in publishing.

In combination with a descriptive reference database, the cryptic form of the REF_CODE can be (and is) attached to a more extensible information listing. For instance, while the REF_CODE carries only the first page number of a reference, the Reference Database carries the first and last page numbers of the article. Obviously, the same qualifications apply to titles and authors which are highly abbreviated in the REF_CODE, but more fully represented in the Database.

The same principles could be used to fully link a REF_CODE to data cubes, CD–ROMs, external databases, animations, simulations, time–tagged data, *etc*. While the Reference Code is compact, it is not yet saturated; there are still fields with room for added pointers to the new directions that the publishing of astronomical data may take in the immediate future.

Acknowledgments

We thank Helmut Abt and the rest of the NED and SIMBAD groups for their help in defining the reference codes. Table 24–2 has been prepared with the kind help of Suzanne Laloë at the Institut d'Astrophysique de Paris. NED is a research support program operated by the Jet Propulsion Laboratory, California Institute of Technology, under contract with the National Aeronautics and Space Administration (Astrophysics Division, Science Operations Branch). SIMBAD is maintained by the Centre de Données astronomiques de Strasbourg, France.

Access Pointers

[→1] http://www.ipac.caltech.edu/NED/refcode.html
[→2] http://cdsweb.u-strasbg.fr/simbad/refcode.html

REF_CODE	Title
A&A..	Astron. Astrophys.
A&AS.	Astron. Astrophys., Suppl. Ser.
Afz..	Astrofizika
AJ...	Astron. J.
ApJ..	Astrophys. J.
ApJS.	Astrophys. J., Suppl. Ser.
ARep.	Astronomy Reports (formerly Soviet Astronomy)
AstL.	Astronomy Letters (formerly Soviet Astronomy Letters)
IAUC.	IAU Circulars.
MNRAS	Monthly Notices of the Royal Astronomical Society
Natur	Nature
PASJ.	Publ. Astron. Soc. Jap.
PASP.	Publ. Astron. Soc. Pac.
SvA..	Soviet Astronomy (now Astronomy Reports)
SvAL.	Soviet Astronomy Letters (now Astronomy Letters)

Table 24–1: NED Core Journal Acronyms

REF_CODE	Title
A&A..	Astronomy & Astrophysics
A&AS.	Astronomy & Astrophysics Suppl. Ser.
A&ARv	Astronomy and Astrophysics Review
AcA..	Acta Astronomica
AcASn	Acta Astronomica Sinica
AcApS	Acta Astrophysica Sinica
ACiCh	Astronomical Circular (Nankin)
Afz..	Astrofizika
AGAb.	Astron. Gesellschaft Abstract Ser.
AISof	Astrophysical Investigations - Sofia
AJ...	Astronomical Journal
AN...	Astronomische Nachrichten
APh..	Astroparticle Physics
ApJ..	Astrophysical Journal
ApJS.	Astrophys. J. Suppl. Ser.
ApL..	Astrophysical Letters and communications
Ap&SS	Astrophys. & Space Science
Aster	Aster
Ast..	Astronomy
ATsir	Astron. Tsirk.
AuJPh	Australian J. Phys.
AZh..	Astronomicheskij Zhurnal
BAAS.	Bull. American Astron. Soc.
BaltA	Baltic Astronomy
BCFHT	Bull. CFHT
BICDS	Bull. Inf. Centre Données Strasbourg
BOBeo	Bull. Obs. Astron. Beograd
C&E..	Ciel et Espace
ChA&A	Chinese Astron. & Astrophys.
C&T..	Ciel et Terre
ComAp	Comments on Astrophysics and space physics
CoSka	Cont. Astron. Obs. Skalnate Pleso
CR2..	Comptes Rendus Acad. Sci. Ser. II
EL...	Europhysics Letters
EM&P.	Earth, Moon, and Planets
FCPh.	Fundamentals of Cosmic Physics
FoPh.	Foundations of Physics
Gemin	GEMINI Newsletter (RGO)

Table 24–2: Journals scanned for SIMBAD bibliography. A full up-to-date list is available on-line via WWW [→2]

REF_CODE	Title
Gri0.	Griffith Obs.
IAUC.	IAU Circular
IBVS.	IAU Inform. Bull. Variable Stars
Icar.	Icarus
IrAJ.	Irish Astron. J.
ISKZ.	Investigations of the Sun and Red Stars
IzKry	Izv. Krym. Astrofiz. Obs.
JApA.	Journal of Astrophysics and Astronomy
JAF..	J. Astronomes Francais
JAVSO	J. American Assoc. Variable Star Obs.
JBAA.	J. British Astron. Ass.
JBIS.	J. British Interplanetary Society
JHA..	Journal for the History of Astronomy
JRASC	J. Royal Astron. Soc. Canada
KFNT.	Kinematika Fizika Nebesnykh Tel. (Kiev)
KoIs.	Kosmic. Issl. (Cosmic research)
LAstr	L'Astronomie
Mercu	Mercury
Msngr	The Messenger
MitAG	Mitt. Astron. Gesellschaft
MitVS	Mitt. Verand. Sterne
MNRAS	Mon. Not. Royal Astron. Soc.
MNSSA	Mon. Not. Astron. Soc. South Africa
MmSAI	Mem. Soc. Astron. Ital.
Natur	Nature
NewSc	New Scientist
Obs..	The Observatory
PASAu	Proc. Astron. Soc. Australia
PASJ.	Publ. Astron. Soc. Japan
PASP.	Publ. Astron. Soc. Pacific
PAZh.	Pis'ma Astron. Zh.
PBei0	Publ. Beijing Astron. Obs.
PhR..	Physics Reports
PhRvA	Physical Review A
PhRvD	Physical Review D
PhRvL	Physical Review Letters
PhS..	Physica Scripta

Table 24–2: Journals scanned for SIMBAD bibliography – cont.

REF_CODE	Title
PhT..	Physics Today
PNAOJ	Publ. Nat. Astron. Obs. Japan
POBeo	Publ. Obs. Astron. Beograd
PPMtO	Publ. Purple Mountain Obs.
P&SS.	Planetary and Space Sciences
PTRSL	Philosophical Transactions R. Soc. London
PZ...	Peremennye Zvezdy (variable stars)
QJRAS	Quart. J. Royal Astron. Soc.
Rech.	La Recherche
RMxAA	Rev. Mexicana Astron. Astrofis.
RoAJ.	Romanian Astronomical Journal
RPPh.	Reports on Progress in Physics
RvPD.	Revue du Palais de la Découverte
SAAOC	South African Astron. Obs. Circ.
SoByu	Soobshch. Byurakan Obs.
Sci..	Science
SciAm	Scientific American
SciN.	Science News
SSRev	Space Science Reviews
S&T..	Sky & Telescope
S&W..	Sterne und Weltraum
VA...	Vistas in Astronomy

Table 24–2: Journals scanned for SIMBAD bibliography – end.

REF_CODE	Title
A&R..	Astron. Raumfahrt
AAfz.	Astrometria Astrofizika
AdAAp	Advances in Astronomy and Astrophysics
AExpr	Astron. Express
AISAO	Astrofizik. Issledovanija, Special Astrophys. Obs.
AnAp.	Ann. Astrophys.
AnCap	Annals Cape Obs.
AnTok	Ann. Tokyo Astron. Obs.
ArA..	Arkiv for Astron.
ARA&A	Annual Review of Astronomy and Astrophysics
BAICz	Bull. Astron. Inst. Czech.
BAN..	Bull. Astron. Inst. Netherlands
BANS.	Bulletin of the Astronomical Institutes of the Netherlands, Supplement Series
BITon	Bol. Inst. Tonantzintla
BOTT.	Bol. Obs. Tonantzintla Tacub.
CAFOE	Circulaire de l'Association Francaise d'Observateurs d'Etoiles Variables
CiBAA	Circular of the British Astronomical Association
CoAsi	Contr. Asiago
CR...	C.R. Acad. Sci.
CRA..	C.R. Acad. Sci. Ser. A
CRB..	C.R. Acad. Sci. Ser. B
ESOSP	European Southern Observatory – Scientific Preprints
IAUCo	IAU Colloquium
IAUS.	IAU Symposium
IAUT.	IAU Transactions
NPhS.	Nature, Physical Science
Orion	Orion
PhRvB	Physical Review B
PhRvC	Physical Review C
Stern	Die Sterne
Urani	Urania

Table 24–3: Some Additional Acronyms. A full list is kept on-line and available via WWW [→1] and [→2]

REF_CODE	Title
1960CGCG..C01....0Z	Zwicky, Herzog, and Wild. 1960, CGCG Vol I
1963CGCG..C02....0Z	Zwicky and Herzog. 1963, CGCC Vol II
1966CGCG..C03....0Z	Zwicky and Herzog. 1966, CGCG Vol III
1968CGCG..C04....0Z	Zwicky and Herzog. 1968, CGCG Vol IV
1965CGCG..C05....0Z	Zwicky, Karpowicz, and Kowal. 1965, CGCG Vol V
1968CGCG..C06....0Z	Zwicky and Kowal. 1968, CGCC Vol VI
1982ESO...C......0L	Lauberts. 1982, ESO/Uppsala Survey
1989ESOLV.C......0L	Lauberts and Valentijn. 1989, ESO–LV Photometry
1962MCG...C01....0V	Vorontsov–Velyaminov and Krasnogorskaja. 1962, MCG Part I
1964MCG...C02....0V	Vorontsov–Velyaminov and Arhipova. 1964, MCG Part II
1963MCG...C03....0V	Vorontsov–Velyaminov and Arhipova. 1963, MCG Part III
1968MCG...C04....0V	Vorontsov–Velyaminov and Arhipova. 1968, MCG Part IV
1974MCG...C05....0V	Vorontsov–Velyaminov and Arhipova. 1974, MCG Part V
1971CGPG.........0Z	Zwicky. 1971, Selected Compact and Post–Eruptive Galaxies
1973UGC...C......0N	Nilson. 1973, UGC
1974UGCA..C......0N	Nilson. 1974, Selected Non–UGC Galaxies
1976RC2...C......0d	de Vaucouleurs, de Vaucouleurs, and Corwin. 1976, RC2
1991RC3...C......0d	de Vaucouleurs, *et al.* . 1991, RC3
1981RSA...C......0S	Sandage and Tammann. 1981, Revised Shapley–Ames
1983RVG...C......0P	Palumbo et al. 1983, Radial Velocity Catalog
1985Q&AN2.C......0V	Veron–C. and Veron. 1985, Quasars and Active Nuclei, 2nd ed
1985SGC...C......0C	Corwin, de Vaucouleurs, and de Vaucouleurs. 1985, Southern Galaxy Catalog
1959VV....C......0V	Vorontsov–Velyaminov. 1959, Interacting Galaxies

Table 24–4: Reference Codes for Selected Catalogs

REF_CODE	Title
CIT..	California Institute of Technology (U.S.A.)
CornU	Cornell University (U.S.A.)
UCSD.	University of California, San Diego (U.S.A.)
UEdin	University of Edinburgh (United Kingdom)
UPari	University of Paris (France)
UTole	University of Toledo (U.S.A.)
UToro	University of Toronto (Canada)
YeshU	Yeshiva University (Russia)

Table 24–5: Examples of University Codes

Authors

Alberto Accomazzi
   Smithsonian Astrophysical Observatory, Cambridge, MA 02138, USA
   alberto@cfa.harvard.edu

Miguel A. Albrecht
   European Southern Observatory, Karl–Schwarzschild–Str. 2, D–85748 Garching, Germany
   malbrech@eso.org

H. Andernach
   IUE Observatory, Villafranca, Apartado 50727, E-28080 Madrid, Spain
   hja@vilspa.esa.es

S. G. Ansari
   ESIS, Information Systems Division of ESA, ESRIN, Via G. Galilei, I-00044 Frascati, Italy
   salim@mail.esrin.esa.it

Paul Barrett
   NASA Goddard Space Flight Center, Compton Observatory Science Support Centre, Greenbelt, MD 20771, USA
   barrett@gsfc.nasa.gov

M. Barylak
   ESA IUE Observatory, P.O.Box 50727, E-28080 Madrid, Spain. Affiliated to the Astrophysics Division, Space Science Department, European Space Agency
   mb@vilspa.esa.es

Judy Bennett
   Infrared Processing and Analysis Center, Jet Propulsion Laboratory, CALTECH, Pasadena, CA 91125, USA
   jdb@tacos.caltech.edu

François Bonnarel
   CDS, Observatoire astronomique de Strasbourg, 11 rue de l'Université, 67000 Strasbourg, France
   bonnarel@astro.u-strasbg.fr

Jean Borsenberger
   Institut d'Astrophysique, 98bis Bld. Arago, 75014 Paris, France
   borsen@iap.fr

L. Bottinelli

Observatoire de Paris-Meudon, 92195 Meudon Principal Cedex, France
`bottin@obspm.fr`

H. Brunner

Astronomisches Institut Tübingen, Waldhäuserstr. 64, 72076 Tübingen, Germany

H.G. Corwin Jr.

Infrared Processing and Analysis Center, Jet Propulsion Laboratory, CALTECH, Pasadena, CA 91125, USA
`hgcjr@ipac.caltech.edu`

Dennis R. Crabtree

Dominion Astrophysical Observatory, Victoria, Canada
`crabtree@dao.nrc.ca`

Michel Crézé

CDS, Observatoire astronomique de Strasbourg, 11 rue de l'Université, 67000 Strasbourg, France
`creze@astro.u-strasbg.fr`

Marlene Cummins

Astronomy Library, University of Toronto, 60 St. George St. Rm 1306, Toronto, Ontario, Canada M5S 1A7
`library@astro.utoronto.ca`

Erik Deul

Leiden Observatory, P.O. Box 9513, NL-2300 RA Leiden, The Netherlands
`deul@strw.leidenuniv.nl`

H. Di Nella

Observatoire de Lyon, 69651 Saint-Genis Laval Cedex, France
`dinella@obs.univ-lyon1.fr`

C. Driessen

Deimos Ltd., Guernsey, UK
`cd@vilspa.esa.es`

Pascal Dubois

CDS, Observatoire astronomique de Strasbourg, 11 rue de l'Université, 67000 Strasbourg, France
`dubois@astro.u-strasbg.fr`

Daniel Durand

Dominion Astrophysical Observatory, Victoria, Canada
`durand@dao.nrc.ca`

N. Durand

Observatoire de Paris-Meudon, 92195 Meudon Principal Cedex, France

R. Ebert

Infrared Processing and Analysis Center, CALTECH, Pasadena CA 91125, USA
`rick@ipac.caltech.edu`

Daniel Egret
: *CDS, Observatoire astronomique de Strasbourg, 11 rue de l'Université, 67000 Strasbourg, France*
egret@astro.u-strasbg.fr

Guenther Eichhorn
: *Smithsonian Astrophysical Observatory, Cambridge, MA 02138, USA*
gei@cfa.harvard.edu

Nicolas Epchtein
: *Observatoire de Paris-Meudon, DESPA, 92195 Meudon Principal CEDEX, France*
epchtein@obspm.fr

R. Garnier
: *Observatoire de Lyon, 69651 Saint-Genis Laval Cedex, France*
garnier@obs.univ-lyon1.fr

Severin Gaudet
: *Dominion Astrophysical Observatory, Victoria, Canada*
gaudet@dao.nrc.ca

Françoise Genova
: *CDS, Observatoire astronomique de Strasbourg, 11 rue de l'Université, 67000 Strasbourg, France*
genova@astro.u-strasbg.fr

Paolo Giommi
: *ESIS, Information Systems Division of ESA, ESRIN, Via G. Galilei, I-00044 Frascati, Italy*
pgiommi@mail.esrin.esa.it

L. Gouguenheim
: *Observatoire de Paris-Meudon, 92195 Meudon Principal Cedex, France*

Daniel E. Harris
: *Smithsonian Astrophysical Observatory, Cambridge, MA 02138 USA*
harris@cfa.harvard.edu

André Heck
: *Strasbourg Astronomical Observatory, 11, rue de l'Université, F-67000 Strasbourg, France*
heck@cdsxb6.u-strasbg.fr, http://cdsweb.u-strasbg.fr/~heck

George Helou
: *Infrared Processing and Analysis Center, Jet Propulsion Laboratory, CALTECH, Pasadena CA 91125, USA*
gxh@ipac.caltech.edu

Norman Hill
: *Dominion Astrophysical Observatory, Victoria, Canada*
nhill@dao.nrc.ca

Jane Holmquist

Astrophysics Library, Princeton University, Princeton NJ 08544, USA
`jane@astro.princeton.edu`

Gérard Jasniewicz

CDS, Observatoire astronomique de Strasbourg, 11 rue de l'Université, 67000 Strasbourg, France
`gerard@astro.u-strasbg.fr`

Michael J. Kurtz

Smithsonian Astrophysical Observatory, Cambridge MA 02138, USA
`kurtz@cfa.harvard.edu`

C. LaGue

Infrared Processing and Analysis Center, Jet Propulsion Laboratory, CALTECH, Pasadena CA 91125, USA
`cher@ipac.caltech.edu`

David Leisawitz

Astrophysics Data Facility, Code 631, NASA/Goddard Space Flight Center, Greenbelt MD 20771, USA
`leisawitz@stars.gsfc.nasa.gov`

Soizick Lesteven

CDS, Observatoire astronomique de Strasbourg, 11 rue de l'Université, 67000 Strasbourg, France
`lesteven@astro.u-strasbg.fr`

Barry F. Madore

Infrared Processing and Analysis Center, Jet Propulsion Laboratory, CALTECH, Pasadena, CA 91125, USA
`barry@ipac.caltech.edu`

M.C. Marthinet

Observatoire de Lyon, 69651 Saint-Genis Laval Cedex, France
`marthinet@obs.univ-lyon1.fr`

John C. Mather

Laboratory for Astronomy and Solar Physics, Code 685, NASA/Goddard Space Flight Center, Greenbelt MD 20771, USA
`mather@stars.gsfc.nasa.gov`

Joe Mazzarella

Infrared Processing and Analysis Center, CALTECH, Pasadena CA 91125, USA
`mazz@ipac.caltech.edu`

Jean-Claude Mermilliod

Institut d'Astronomie, Université de Lausanne, 51 ch. des Maillettes CH-1290 Chavannes-des-Bois, Switzerland
`mermio@scsun.unige.ch`

Uta Michold

European Southern Observatory, Karl-Schwarzschild-Str. 2, D-85748 Garching, Germany
esolib@eso.org

Stephen C. Morris

Dominion Astrophysical Observatory, Victoria, Canada
morris@dao.nrc.ca

Stephen S. Murray

Smithsonian Astrophysical Observatory, Cambridge MA 02138, USA
ssm@cfa.harvard.edu

Fionn Murtagh

Space Telescope – European Coordinating Facility, European Southern Observatory, Karl-Schwarzschild-Str. 2, D–85748 Garching, Germany. Affiliated to the Astrphysiscs Division, Space Science Department, European Space Agency.
fmurtagh@eso.org

François Ochsenbein

CDS, Observatoire astronomique de Strasbourg, 11 rue de l'Université, 67000 Strasbourg, France
francois@astro.u-strasbg.fr

G. Paturel

Observatoire de Lyon, 69651 Saint-Genis Laval Cedex, France
paturel@obs.univ-lyon1.fr

C. Petit

Observatoire de Lyon, 69651 Saint-Genis Laval Cedex, France

Benoît Pirenne

Space Telescope – European Coordinating Facility, European Southern Observatory, Karl-Schwarzschild-Str. 2, D–85748 Garching, Germany
bpirenne@eso.org

R. Pisarski

NASA Goddard Space Flight Center, Code 631, Green Belt, Maryland 20771, USA
pisarski@nssdca.gsfc.nasa.gov

J. D. Ponz

ESA ECNOD/VCS, P.O.Box 50727, E-28080 Madrid, Spain
jdp@v3300.vilspa.esa.es

Michèle Péron

European Southern Observatory, Karl–Schwarzschild–Str. 2, D–85748 Garching, Germany
mperon@eso.org

Marion Schmitz

Infrared Processing and Analysis Center, Jet Propulsion Laboratory, CALTECH, Pasadena CA 91125, USA
mschmitz@ipac.caltech.edu

Robyn Shobbrook

   Anglo-Australian Observatory, P.O.Box 296, Epping, NSW 2121, Australia
   LIB@aaoepp.oz.au

C. Stern Grant

   Smithsonian Astrophysical Observatory, Cambridge, MA 02138 USA
   stern@cfa.harvard.edu

H. Sun

   Infrared Processing and Analysis Center, Jet Propulsion Laboratory, CALTECH, Pasadena
   CA 91125, USA

A. Talavera

   ESA IUE Observatory, P.O.Box 50727, E-28080 Madrid, Spain. Affiliated to the Astrophysics Division, Space Science Department, European Space Agency
   at@vilspa.esa.es

I. Vauglin

   Observatoire de Lyon, 69651 Saint-Genis Laval Cedex, France
   vauglin@obs.univ-lyon1.fr

W. Voges

   Max-Planck-Institut für Extraterrestrische Physik, P.O. Box 1603, 85740 Garching, Germany

W. Wamsteker

   ESA IUE Observatory, P.O.Box 50727, E-28080 Madrid, Spain. Affiliated to the Astrophysics Division, Space Science Department, European Space Agency
   ww@vilspa.esa.es

Joyce M. Watson

   Smithsonian Astrophysical Observatory, 60 Garden Street MS-12, Cambridge MA 02138,
   USA
   jwatson@cfa.harvard.edu

M. G. Watson

   University of Leicester, Dept. of Physics, University Road, Leicester LE1 4RH, UK
   mgw@star.le.ac.uk

Marc Wenger

   CDS, Observatoire astronomique de Strasbourg, 11 rue de l'Université, 67000 Strasbourg,
   France
   wenger@astro.u-strasbg.fr

Nicholas E. White

   HEASARC, Goddard Space Flight Center, Greenbelt, MD 20771, USA
   white@lheavx.gsfc.nasa.gov

Xiuqin Wu

   Infrared Processing and Analysis Center, Jet Propulsion Laboratory, CALTECH, Pasadena
   CA 91125, USA
   xiuqin@ipac.caltech.edu

H. U. Zimmermann
Max-Planck-Institut für Extraterrestrische Physik, P.O. Box 1603, 85740 Garching, Germany
zim@mpe-garching.mpg.de

Acronyms

2MASS	Two Micron All-Sky Survey
AAS	American Astronomical Society
ACM	Association for Computational Machinery
ADC	Astronomical Data Center
ADS	Astrophysics Data System
AIT	Astronomisches Institut Tübingen
ALD	Astronomy from Large Databases Conference
ANSI/NISO	American National Standards Institute/ National Information Standards Organization
API	Applications Program Interface
ASIAS	Astrophysics Science Information and Abstract Service
AXAF	Advanced X-ray Astronomy Facility
BATSE	GRO Burst and Transient Source Experiment
BDA	Base de Donnees des Amas (Database for Galactic Open Clusters)
CADC	Canadian Astronomy Data Centre
CALIB*	DENIS Calibration star database
CCDM	Catalogue des Composantes d'Etoiles Doubles et Multiples
CCI	Common Client Interface
CDS	Centre de Données astronomiques de Strasbourg
CDS	Strasbourg astronomical Data Center
CEA	Center for EUV Astrophysic
CFHT	Canada France Hawaii Telescope
CNIDR	Clearinghouse for Networked Information Discovery and Retrieval
CNRS	Centre National de la Recherche Scientifique
COBE	Cosmic Background Explorer
CfA	Center for Astrophysics
DADS	Data Archive and Distribution System
DAO	Dominion Astrophysical Observatory
DENIS	DEep Near Infrared Survey of the Southern Sky
DICB	ESO Data Interface Control Board
DIRBE	COBE Diffuse Infrared Background Experiment

279

DMR	COBE Differential Microwave Radiometer
ECHT	European Conference on Hypertext
EGRET	GRO Energetic Gamma Ray Experiment Telescope
EOLS	The Einstein On-Line Service
ESDB	DENIS Extended Source Database
ESIS	European Space Information System
ESO	European Southern Observatory
EUVE	Extreme Ultraviolet Explorer
EXSAS	ROSAT Extended Scientific Analysis System
FAQ	Frequently Asked Questions
FGDC	Federal Geospatial Data Clearinghouse
FIRAS	COBE Far Infrared Absolute Spectrophotometer
FOURBI	DENIS Follow–Up Relational Base for Images
GIS	geographic information systems
GRO	Compton Gamma-Ray Observatory
GSOC	German Space Operation Center
Glimpse	GLobal IMPlicit SEarch
HEASARC	High Energy Astrophysics Science Archive Center
HRI	High Resolution Imager
HST	Hubble Space Telescope
http	hypertext transfer protocol
IAU	International Astronomical Union
IDL	Interactive Data Language
IPAC	Infrared Processing and Analysis Center
IRAS	Infrared Astronomical Satellite
ISDN	Integrated Services Data Network
ISO	Infrared Space Observatory
ISO	International Organization for Standardization
IUE	International Ultraviolet Explorer
IUESIPS	IUE Spectral Image Processing System
JAM	JYACC Application Manager
JCMT	James Clerk Maxwell Telescope
LDAC	Leiden Data Analysis Center
LEDA	Lyon-Meudon Extragalactic Database
LHEA	Laboratory for High Energy Astrophysics
LODA	DENIS Large Object DAtabase
MBONE	Multicast BackbONE
MIME	Multipurpose Internet Mail Extensions
MIPS	Mission Information and Planning System
MPE	Max-Planck-Institut für Extraterrestrische Physik

NASA	US National Aeronautics and Space Administration
NASA/STI	NASA Scientific and Technical Information
NCSA	National Center for Supercomputing Applications
NED	NASA/IPAC Extragalactic Database
NTT	ESO New Technology Telescope
OPTID	IRAS Optical Identification
OSSE	Oriented Scintillation Spectroscopy Experiment
OST	Observation Summary Table
PASP	Publications of the Astronomical Society of the Pacific
PDAC	Paris Data Analysis Center
PPM	Positions and Proper Motions Catalog
PROFDA	DENIS Processed Frame Database
PROS	Post Reduction Offline System
PSC	IRAS Point Source Catalog
PSS	Palomar Sky Survey
RAL	Rutherford Appleton Laboratory
RDA	ROSAT Data Archive
RDF	Rationalized Data Format
ROSAT	Röntgen Satellite
RRA	ROSAT Result Archive
RSDC	ROSAT Science Data Centres
SASS	Standard Analysis Software System
SERC	UK Science and Engineering Research Council
SIAM	Society for Industrial and Applied Mathematics
SIMBAD	Set of Identifications, Measurements and Bibliography for Astronomical Data
SIRTF	Space Infra-Red Telescope Facility
SODA	DENIS Small Objects DAtabase
SQL	Structured Query Language
STScI	Space Telescope Science Institute
ST–ECF	Space Telescope European Coordinating Facility
STSDAS	Space Telescope Scientific Data Analysis System
SVJPS	Springer-Verlag Journals Preview Service
TMSS	Two Micron Sky Survey
UKRDAC	UK ROSAT Data Archive Centre
ULAS	Union List of Astronomy Serials
ULDA	IUE Uniform Dispersion Low Dispersion
URL	Uniform Resource Locator
URN	Uniform Resource Name
USSP	ULDA Support Software Package

VLT	ESO Very Large telescope
WF/PC	Wide Field/Planetary Camera
WFC	Wide Field Camera
WIRE	Wide Field Infrared Explorer
WWW	World-Wide Web
XRT	X-ray telescope
ZIG	Z39.50 Implementors Group

Index

WF/PC, 240
WFC, 27, 28
white pages service, 240, *see also*
          directories
WHOIS, 239
Wide Field Camera, *see* WFC
Wide Field/Planetary Camera, *see*
          WF/PC
Wide–Field Infrared Explorer, *see*
          WIRE
WIRE, 147
World-Wide Web, *see* WWW
WORM, 52, *see also* optical disk or
          storage technology, 154, 155,
          247
WWW, 3, 32, 82, 110, 185, 235

X-ray spectra, 189
X-ray telescope, *see* XRT
X-Windows, 107
X.500, 239
XANADU, 142
XIMAGE, 188
XRONOS, 190
XRT, 27
Xsimbad, 167
XSPEC, 189
XTE, 139
XUV, 27

yellow pages service, 6, *see also*
          directories, 170, 196

Z39.50, 238
Z39.50 Implementors Group, *see* ZIG
ZDist, 238
ZIG, 238